Computational Fourier Optics
A MATLAB® Tutorial

Tutorial Texts Series

- *Modeling the Imaging Chain of Digital Cameras*, Robert D. Fiete, Vol. TT92
- *Cells Illuminated: In Vivo Optical Imaging*, Lubov Brovko, Vol. TT91
- *Polarization of Light with Applications in Optical Fibers*, Arun Kumar, Ajoy Ghatak, Vol. TT90
- *Computational Fourier Optics: A MATLAB® Tutorial*, David Voelz, Vol. TT89
- *Optical Design of Microscopes*, George Seward, Vol. TT88
- *Analysis and Evaluation of Sampled Imaging Systems*, Richard H. Vollmerhausen, Donald A. Reago, Ronald Driggers, Vol. TT87
- *Nanotechnology: A Crash Course*, Raúl J. Martin-Palma and Akhlesh Lakhtakia, Vol. TT86
- *Direct Detection LADAR Systems*, Richard Richmond, Stephen Cain, Vol. TT85
- *Optical Design: Applying the Fundamentals*, Max J. Riedl, Vol. TT84
- *Infrared Optics and Zoom Lenses, Second Edition*, Allen Mann, Vol. TT83
- *Optical Engineering Fundamentals, Second Edition*, Bruce H. Walker, Vol. TT82
- *Fundamentals of Polarimetric Remote Sensing*, John Schott, Vol. TT81
- *The Design of Plastic Optical Systems*, Michael P. Schaub, Vol. TT80
- *Fundamentals of Photonics*, Chandra Roychoudhuri, Vol. TT79
- *Radiation Thermometry: Fundamentals and Applications in the Petrochemical Industry*, Peter Saunders, Vol. TT78
- *Matrix Methods for Optical Layout*, Gerhard Kloos, Vol. TT77
- *Fundamentals of Infrared Detector Materials*, Michael A. Kinch, Vol. TT76
- *Practical Applications of Infrared Thermal Sensing and Imaging Equipment, Third Edition*, Herbert Kaplan, Vol. TT75
- *Bioluminescence for Food and Environmental Microbiological Safety*, Lubov Brovko, Vol. TT74
- *Introduction to Image Stabilization*, Scott W. Teare, Sergio R. Restaino, Vol. TT73
- *Logic-based Nonlinear Image Processing*, Stephen Marshall, Vol. TT72
- *The Physics and Engineering of Solid State Lasers*, Yehoshua Kalisky, Vol. TT71
- *Thermal Infrared Characterization of Ground Targets and Backgrounds, Second Edition*, Pieter A. Jacobs, Vol. TT70
- *Introduction to Confocal Fluorescence Microscopy*, Michiel Müller, Vol. TT69
- *Artificial Neural Networks: An Introduction*, Kevin L. Priddy and Paul E. Keller, Vol. TT68
- *Basics of Code Division Multiple Access (CDMA)*, Raghuveer Rao and Sohail Dianat, Vol. TT67
- *Optical Imaging in Projection Microlithography*, Alfred Kwok-Kit Wong, Vol. TT66
- *Metrics for High-Quality Specular Surfaces*, Lionel R. Baker, Vol. TT65
- *Field Mathematics for Electromagnetics, Photonics, and Materials Science*, Bernard Maxum, Vol. TT64
- *High-Fidelity Medical Imaging Displays*, Aldo Badano, Michael J. Flynn, and Jerzy Kanicki, Vol. TT63
- *Diffractive Optics–Design, Fabrication, and Test*, Donald C. O'Shea, Thomas J. Suleski, Alan D. Kathman, and Dennis W. Prather, Vol. TT62
- *Fourier-Transform Spectroscopy Instrumentation Engineering*, Vidi Saptari, Vol. TT61
- *The Power- and Energy-Handling Capability of Optical Materials, Components, and Systems*, Roger M. Wood, Vol. TT60
- *Hands-on Morphological Image Processing*, Edward R. Dougherty, Roberto A. Lotufo, Vol. TT59
- *Integrated Optomechanical Analysis*, Keith B. Doyle, Victor L. Genberg, Gregory J. Michels, Vol. TT58
- *Thin-Film Design: Modulated Thickness and Other Stopband Design Methods*, Bruce Perilloux, Vol. TT57
- *Optische Grundlagen für Infrarotsysteme*, Max J. Riedl, Vol. TT56
- *An Engineering Introduction to Biotechnology*, J. Patrick Fitch, Vol. TT55
- *Image Performance in CRT Displays*, Kenneth Compton, Vol. TT54
- *Introduction to Laser Diode-Pumped Solid State Lasers*, Richard Scheps, Vol. TT53

(For a complete list of Tutorial Texts, see http://spie.org/x651.xml.)

Computational Fourier Optics
A MATLAB® Tutorial

David Voelz

Tutorial Texts in Optical Engineering
Volume TT89

Bellingham, Washington USA

Voelz, David George, 1959-
  Computational fourier optics : a MATLAB tutorial / David G. Voelz.
     p. cm.
  Includes bibliographical references and index.
  ISBN 978-0-8194-8204-4
  1. Image processing--Digital techniques--Mathematics. 2. Fourier transform optics. 3. MATLAB. I. Title.
   TA1637.V674 2010
   621.3601'515723--dc22
                                    2010050505

Published by

SPIE
P.O. Box 10
Bellingham, Washington 98227-0010 USA
Phone: +1 360.676.3290
Fax: +1 360.647.1445
Email: Books@spie.org
Web: http://spie.org

Copyright © 2011 Society of Photo-Optical Instrumentation Engineers (SPIE)

All rights reserved. No part of this publication may be reproduced or distributed in any form or by any means without written permission of the publisher.

The content of this book reflects the work and thoughts of the author(s).
Every effort has been made to publish reliable and accurate information herein, but the publisher is not responsible for the validity of the information or for any outcomes resulting from reliance thereon.

Printed in the United States of America.

Fourth Printing

To my dad, my best friend who always wanted to write a book.

Introduction to the Series

Since its inception in 1989, the Tutorial Texts (TT) series has grown to cover many diverse fields of science and engineering. The initial idea for the series was to make material presented in SPIE short courses available to those who could not attend and to provide a reference text for those who could. Thus, many of the texts in this series are generated by augmenting course notes with descriptive text that further illuminates the subject. In this way, the TT becomes an excellent stand-alone reference that finds a much wider audience than only short course attendees.

Tutorial Texts have grown in popularity and in the scope of material covered since 1989. They no longer necessarily stem from short courses; rather, they are often generated independently by experts in the field. They are popular because they provide a ready reference to those wishing to learn about emerging technologies or the latest information within their field. The topics within the series have grown from the initial areas of geometrical optics, optical detectors, and image processing to include the emerging fields of nanotechnology, biomedical optics, fiber optics, and laser technologies. Authors contributing to the TT series are instructed to provide introductory material so that those new to the field may use the book as a starting point to get a basic grasp of the material. It is hoped that some readers may develop sufficient interest to take a short course by the author or pursue further research in more advanced books to delve deeper into the subject.

The books in this series are distinguished from other technical monographs and textbooks in the way in which the material is presented. In keeping with the tutorial nature of the series, there is an emphasis on the use of graphical and illustrative material to better elucidate basic and advanced concepts. There is also heavy use of tabular reference data and numerous examples to further explain the concepts presented. The publishing time for the books is kept to a minimum so that the books will be as timely and up-to-date as possible. Furthermore, these introductory books are competitively priced compared to more traditional books on the same subject.

When a proposal for a text is received, each proposal is evaluated to determine the relevance of the proposed topic. This initial reviewing process has been very helpful to authors in identifying, early in the writing process, the need for additional material or other changes in approach that would serve to strengthen the text. Once a manuscript is completed, it is peer reviewed to ensure that chapters communicate accurately the essential ingredients of the science and technologies under discussion.

It is my goal to maintain the style and quality of books in the series and to further expand the topic areas to include new emerging fields as they become of interest to our reading audience.

James A. Harrington
Rutgers University

Contents

Preface xiii

Chapter 1 Analytic Fourier Theory Review 1

    1.1 A Little History and Purpose 1
    1.2 The Realm of Computational Fourier Optics 2
    1.3 Fourier Transform Definitions and Existence 3
    1.4 Theorems and Separability 3
    1.5 Basic Functions and Transforms 5
    1.6 Linear and Space-Invariant Systems 7
    1.7 Exercises 10
    1.8 References 12

Chapter 2 Sampled Functions and the Discrete Fourier Transform 13

    2.1 Sampling and the Shannon–Nyquist Sampling Theorem 13
    2.2 Effective Bandwidth 15
    2.3 Discrete Fourier Transform from the Continuous Transform 18
    2.4 Coordinates, Indexing, Centering, and Shifting 20
    2.5 Periodic Extension 21
    2.6 Periodic Convolution 24
    2.7 Exercises 26
    2.8 References 27

Chapter 3 MATLAB Programming of Functions, Vectors, Arrays, and Fourier Transforms 29

    3.1 Defining Functions 29
    3.2 Creating Vectors 32
    3.3 Shift for FFT 34
    3.4 Computing the FFT and Displaying Results 36
    3.5 Comparison with Analytic Results 38
    3.6 Convolution Example 39
    3.7 Two Dimensions 41
    3.8 Miscellaneous Hints 43
    3.9 Exercises 45

Chapter 4 Scalar Diffraction and Propagation Solutions ... 47

4.1 Scalar Diffraction ... 47
4.2 Monochromatic Fields and Irradiance ... 48
4.3 Optical Path Length and Field Phase Representation ... 50
4.4 Analytic Diffraction Solutions ... 51
    4.4.1 Rayleigh–Sommerfeld solution I ... 51
    4.4.2 Fresnel approximation ... 53
    4.4.3 Fraunhofer approximation ... 55
4.5 Fraunhofer Diffraction Example ... 56
4.6 Exercises ... 59
4.7 References ... 61

Chapter 5 Propagation Simulation ... 63

5.1 Fresnel Transfer Function (TF) Propagator ... 63
5.2 Fresnel Impulse Response (IR) Propagator ... 64
5.3 Square Beam Example ... 66
5.4 Fresnel Propagation Sampling ... 69
    5.4.1 Square beam example results and artifacts ... 69
    5.4.2 Sampling regimes and criteria ... 72
    5.4.3 Criteria applied to square beam example ... 74
    5.4.4 Propagator accuracy ... 75
    5.4.5 Sampling decisions ... 77
    5.4.6 Split-step simulation, windowing, and expanding grids ... 78
5.5 Fraunhofer Propagation ... 79
5.6 Coding Efficiency ... 83
5.7 Exercises ... 83
5.8 References ... 86

Chapter 6 Transmittance Functions, Lenses, and Gratings ... 89

6.1 Tilt ... 89
6.2 Focus ... 93
6.3 Lens ... 96
6.4 Gratings and Periodic Functions ... 98
    6.4.1 Cosine magnitude example ... 99
    6.4.2 Square-wave magnitude example ... 102
    6.4.3 One-dimensional model ... 105
    6.4.4 Periodic model ... 106
6.5 Exercises ... 108
6.6 References ... 111

Chapter 7 Imaging and Diffraction-Limited Imaging Simulation ... 113

7.1 Geometrical Imaging Concepts ... 113
7.2 Coherent Imaging ... 116

  7.2.1 Coherent imaging theory 116
  7.2.2 Coherent transfer function examples 117
  7.2.3 Diffraction-limited coherent imaging simulation 119
  7.2.4 Rough object 124
 7.3 Incoherent Imaging 127
  7.3.1 Incoherent imaging theory 127
  7.3.2 Optical transfer function examples 128
  7.3.3 Diffraction-limited incoherent imaging simulation 129
 7.4 Exercises 132
 7.5 References 139

Chapter 8 Wavefront Aberrations 141

 8.1 Wavefront Optical Path Difference 141
 8.2 Seidel Polynomials 142
  8.2.1 Definition and primary aberrations 142
  8.2.2 MATLAB function 144
 8.3 Pupil and Transfer Functions 146
  8.3.1 Pupil function 146
  8.3.2 Imaging transfer functions 147
 8.4 Image Quality 147
  8.4.1 Point spread function 147
  8.4.2 Modulation transfer function 148
 8.5 Lens Example—PSF and MTF 148
 8.6 Wavefront Sampling 153
 8.7 Superposition Imaging Example 157
  8.7.1 Image plane PSF map 157
  8.7.2 Image simulation 160
  8.7.3 Practical image simulation 163
 8.8 Exercises 163
 8.9 References 168

Chapter 9 Partial Coherence Simulation 169

 9.1 Partial Temporal Coherence 170
  9.1.1 Quasi-monochromatic light 170
  9.1.2 Partial temporal coherence simulation approach 172
  9.1.3 Partial temporal coherence example 173
 9.2 Partial Spatial Coherence 177
  9.2.1 Stochastic transmittance screen 177
  9.2.2 Partial spatial coherence simulation approach 178
  9.2.3 Partial spatial coherence example 182
 9.3 Reducibility, Number of Spectral Components, and Phase Screens 186
 9.4 Exercises 187
 9.5 References 189

Appendix A Fresnel Propagator Chirp Sampling 191

A.1 Fresnel Transfer Function Sampling 191
    A.1.1 Oversampled transfer function 192
    A.1.2 Critically sampled transfer function 194
    A.1.3 Undersampled transfer function 194
A.2 Fresnel Impulse Response Function Sampling 195
    A.2.1 Undersampled impulse response 196
    A.2.2 Critically sampled impulse response 196
    A.2.3 Oversampled impulse response 197
A.3 Summary 198
A.4 References 198

Appendix B Fresnel Two-Step Propagator 199

B.1 Approach 199
B.2 Sampling Considerations 202
    B.2.1 Similar side lengths 203
    B.2.2 Significantly different side lengths 203
    B.2.3 Comments and recommendations 204
B.3 MATLAB Code 204
B.4 References 205

Appendix C MATLAB Function Listings 207

C.1 Circle 207
C.2 Jinc 207
C.3 Rectangle 207
C.4 Triangle 208
C.5 Unit Sample "Comb" 208
C.6 Unit Sample "Delta" 208

Appendix D Exercise Answers and Results 209

D.1 Chapter 1 209
D.2 Chapter 2 210
D.3 Chapter 3 211
D.4 Chapter 4 212
D.5 Chapter 5 214
D.6 Chapter 6 217
D.7 Chapter 7 220
D.8 Chapter 8 223
D.9 Chapter 9 225

Index 229

Preface

This book began as a collection of notes and computer examples prepared for a first-year graduate course on Fourier optics. In teaching Fourier optics over a number of years, I found that I developed a better conceptual understanding of the analytic material after setting up examples for the class on the computer. The examples required careful consideration of the sample coordinates, amplitude scaling, practical dimensions, display settings, sampling conditions, and a number of other issues. It wasn't long before I started designing computer exercises for the students to do—figuring that if it helped me, it would probably help them. In addition, applying the theory to produce a display of a beam pattern or a blurry image of some object seemed to bring the application of Fourier optics to life for many students.

At the same time, the research being performed by my group at New Mexico State University involved wave optics simulation of laser beam propagation through atmospheric turbulence. The synergy of the teaching and research activities led to the idea of a book on computer methods and Fourier optics. I did some research and found a scattering of material on numerical Fourier optics, but no book with the content I envisioned. So with that, the project began.

Computational Fourier Optics is a text that shows the reader in a tutorial form how to implement Fourier optical theory and analytic methods on the computer. A primary objective is to give students of Fourier optics the capability of programming their own basic wave optic beam propagations and imaging simulations. The book will also be of interest to professional engineers and physicists learning Fourier optics simulation techniques—either as a self-study text or a text for a short course. For more advanced study, the latter chapters and appendices provide methods and examples for modeling beams and pupil functions with more complicated structure, aberrations, and partial coherence.

For a student in a course on Fourier optics, I envision this book as a companion to any of several excellent textbooks on Fourier optical theory. I felt a companion book should be concise, accessible, and practical—so those are also goals for this text.

The book begins in Chapter 1 with a short review of the Fourier optical results that are central to wave optics simulation development. The review is intended to be a quick, consolidated reference.

In Chapter 2 the discrete Fourier transform (DFT) is developed, of which the fast Fourier transform (FFT) version is a primary tool for simulations. FFT scaling aspects, index formatting, and other differences from the analytic transform are introduced. These differences later come to play in the scaling and interpretation of the simulation results.

The hands-on tutorial part of the book begins in Chapter 3 where step-by-step examples are presented that involve programming functions, vectors, equations, and taking transforms in MATLAB®. Students with a range of backgrounds—electrical engineers, astronomers, physicists—take my Fourier optics course. The non-engineers often have never used MATLAB, so the idea of combining a MATLAB tutorial with a computational Fourier optics tutorial was natural and led to Chapter 3. The MATLAB programming environment is optimized for vector and matrix operations; therefore, it is a good tool for Fourier optics simulation, which generally involves at least two dimensions. Furthermore, MATLAB has a heritage in this subject since several optical propagation codes, such as the AOTools and WaveProp toolboxes, are written in MATLAB. The material in this chapter has been tested by students in my Fourier optics course, and even those without any MATLAB experience have found they could get up and going quickly with the tutorial.

Chapter 4 is a quick review and summary of scalar diffraction and optical propagation theory. The expressions presented in Chapter 4 are taken into the computer domain in Chapter 5. Implementations of the Fresnel and Fraunhofer diffraction expressions are described with step-by-step coding instructions. The methods are demonstrated for an illuminated aperture. Attention is paid to sampling issues that can be the bane of wave optics propagation simulations.

Chapter 6 covers techniques that add further application to the diffraction simulations. Methods are described for applying tilt and focus to an optical wavefront, and lenses and diffraction gratings are considered.

A review of coherent and incoherent imaging theory and modeling techniques applied to diffraction-limited imaging examples are presented in Chapter 7. Imaging simulation is extended in Chapter 8 to the more practical circumstance involving wavefront aberrations.

Chapter 9 provides a short review of coherence theory and demonstrates approaches for simulating partial temporal and partial spatial coherent illumination.

Exercises at the end of each chapter (with answers in the back of the book) give the reader a chance to work with both theory and computer implementations.

The appendices cover: (a) further sampling details for Fresnel diffraction; (b) a two-step diffraction propagation technique that allows arbitrary grid scaling between the source and observation planes; (c) listings of basic MATLAB functions developed in the text; and (d) answers to the exercises.

Please visit http://www.ece.nmsu.edu/~davvoelz/cfo/ for updates, errata, files and other resources.

This book would have never happened were it not for a sabbatical leave in 2008 in the Upper Peninsula of Michigan. I owe Mike Roggemann at Michigan Tech a big debt for all of his care and feeding of a displaced New Mexican. My discussions with him, on and off the lake, helped shape much of the content of this book. I also thank the faculty and staff at Michigan Tech for all their support. Go Huskies!

As this project was getting underway, Jason Schmidt kindly sent a first draft of his book *Numerical Simulation of Optical Wave Propagation with Examples in MATLAB®*. I tried to avoid studying it too closely as I wanted to put my own spin on related material. But I had to peek from time to time to see what he had to say on certain matters. His book was a valuable resource.

Xifeng Xiao at New Mexico State University deserves credit for pioneering much of the partial coherence material. She also combed through all the chapters, working examples and checking equations. Our discussions over the years on numerical simulation are deeply imbedded in this book. It has been a great pleasure to work with her.

The students of a succession of Fourier optics courses since 2003 at New Mexico State University have been, often unknowingly, a constant source of insight and inspiration for this book. Their reactions and feedback to the material helped change many things for the better and encouraged me to keep going.

For all those spur-of-the-moment questions and sudden inquiries of how-does-that-work, I thank my colleagues at the Klipsch School of ECE at New Mexico State University, especially Deva Borah, Laura Boucheron, Chuck Creusere, Philip DeLeon and Mike Giles - a good group of folks.

Finally I cannot thank my wife, Judi, enough for supporting this project in every way, including proofreading the manuscript. Our children, Alex, Katie and Brian, have had to deal with an absent dad while I worked on this book, so I thank them for their patience. My family is my support and I couldn't do what I do without them!

David Voelz
December 2010

Chapter 1
Analytic Fourier Theory Review

1.1 A Little History and Purpose

The branch of optical science known today as "Fourier optics" had its genesis in the 1940s through the 1960s with the application of new telecommunications and circuit design analysis techniques in optical diffraction theory.[1] In 1968 this upstart discipline was given a permanent foothold with the publication of *Introduction to Fourier Optics*, by Joseph W. Goodman, a seminal textbook that explained and united the fundamental concepts, and which continues to add significantly to the application of Fourier optics in subsequent editions.[2] Fourier optics is now the cornerstone for the analysis of diffraction, coherence, and imaging, as well as specialized topics such as wavefront control, propagation through random media, and holography.

The study of Fourier optics today leads naturally toward the computer for at least two reasons: (1) diffraction integral expressions are difficult to solve analytically for all but a few of the simplest aperture functions, and (2) the fast Fourier transform (FFT) algorithm combined with the linear systems framework of Fourier optics provides an extremely efficient computational approach for solving wave optics problems.

Certainly, the computer can be applied directly in finding exceedingly accurate solutions to diffraction problems using numerical integration techniques.[3] However, this book is really about the FFT and how to apply it to a variety of Fourier optics problems. The computer coding steps mirror the analytic concepts and the FFT's speed makes it possible to perform thousands of optical propagation or imaging simulations in a reasonable amount of time. In fact, the methods explored in this book form the basis for wave (or physical) optics simulation tools that are widely used in industry. But, of course, there's no free lunch (…if there were, perhaps we could be eating *while* studying Fourier optics…). It turns out the FFT is an accomplice to various numerical artifacts. We do our best in this book to expose these issues and provide constraints to help minimize the damage.

This is also a tutorial text with step-by-step instructions, not only for coding Fourier optics problems, but also for MATLAB, our software application of choice. So, if you are new to MATLAB, don't worry! Chapter 3 starts at the beginning ("Open MATLAB") and leads you through the basics of working with the FFT. By the end of the book you will be programming diffraction problems involving partially coherent light—at least that's the goal! Exercises at the end of the chapters give you room to tinker with the programs and stretch out with your own code.

It is assumed the reader has some familiarity with Fourier optics. Presenting the topic from the ground up is too much material to cover and would obscure our purpose. However, the analytic theory required is presented in summary form throughout the text. The notation and form closely follow Goodman's presentation in *Introduction to Fourier Optics*.[2] For further details and explanations of the analytic foundations of Fourier theory and Fourier optics the reader is encouraged to consult Goodman's book as well as the many other excellent references that exist on the topic.[4-7]

1.2 The Realm of Computational Fourier Optics

In this book, the variables, vectors, and arrays in the computer code are defined as much as possible in terms of physical quantities. For example, the coordinates of samples in an array that models a spatial plane are defined in units of meters. Integers for indexing arrays show up only when they can't be avoided. This approach allows a clear connection between the physical world being modeled and the computer code. MATLAB's vectorized structure is also suited to this approach. Thus, programming examples presented in the book involve specific aperture sizes, wavelengths, and distances. Although some examples are simply academic, others are something one might encounter in the real world. However, the reader will soon notice an emergent theme: the finite size of the sample array in the computer limits the range of parameters that can be considered.

We might consider this difficulty in light of the optical designer's dilemma: When does one transition between a geometrical optics prediction of system performance and a wave optics prediction? The difference between these predictors is that geometrical optics assumes rectilinear (straight-line) propagation of the rays of light and ignores diffractive spreading due to the wave nature of light. The usual answer for the dilemma is that for small departures from perfection (near the "diffraction limit") a wave optics description is needed. For large departures a geometrical ray optics description, which has more flexible implementation options, is adequate.[8,9]

So, although analytic Fourier optics theory is quite general, the finite array size tends to limit the computer modeling to the "near-perfection" situations. Typically, this means small divergence angles for optical beam propagation, small simulated image area, and so forth. For practical applications, this is the same realm as the wave optics performance prediction for optical system design.

The remainder of this chapter is a summary of the fundamental Fourier transform definitions, theorems, basic functions, and transform pairs. A review of linear systems theory is also included. So, let's go!

1.3 Fourier Transform Definitions and Existence

Fourier optics problems often involve two spatial dimensions. The analytic Fourier transform of a function g of two variables x and y is given by

$$G(f_X, f_Y) = \int\int_{-\infty}^{\infty} g(x,y) \exp\left[-j2\pi(f_X x + f_Y y)\right] dx\,dy, \qquad (1.1)$$

where $G(f_X, f_Y)$ is the transform result and f_X and f_Y are independent frequency variables associated with x and y, respectively. This operation is often described in a shorthand manner as $\Im\{g(x,y)\} = G(f_X, f_Y)$. Similarly, the analytic inverse Fourier transform is given by

$$g(x,y) = \int\int_{-\infty}^{\infty} G(f_X, f_Y) \exp\left[j2\pi(f_X x + f_Y y)\right] df_X\,df_Y. \qquad (1.2)$$

The shorthand notation for this operation is $\Im^{-1}\{G(f_X, f_Y)\} = g(x,y)$.

For the Fourier transform to be realizable in a mathematical sense, $g(x,y)$ must satisfy certain sufficient conditions. These conditions are commonly listed as:

(a) g must be absolutely integrable over the infinite range of x and y;

(b) g must have only a finite number of discontinuities; and

(c) g must have no infinite discontinuities.

Goodman[2] illustrates that in a number of important cases, one or more of these conditions can be weakened, and a generalized transform approach using idealized mathematical functions can be employed to find useful transform representations. Some generalized transform results of interest include

$$\Im\{1\} = \delta(f_X, f_Y),$$

$$\Im\{\cos(2\pi f_0 x)\} = \tfrac{1}{2}\delta(f_X - f_0, f_Y) + \tfrac{1}{2}\delta(f_X + f_0, f_Y),$$

where δ is the Dirac delta function.

1.4 Theorems and Separability

The theorems listed in Table 1.1 find considerable application in Fourier analysis. In Table 1.1, A, B, a, and b are scalar constants.

An important property of certain functions is *separability*. A two-dimensional (2D) function is separable if it can be written as the product of two functions of a single variable, such as

$$g_S(x, y) = g_X(x)g_Y(y). \tag{1.3}$$

Separability reduces the Fourier transform of a 2D function to the product of two one-dimensional (1D) transforms or

$$\Im\{g_S(x, y)\} = \Im\{g_X(x)\}\Im\{g_Y(y)\}. \tag{1.4}$$

Table 1.1 Fourier transform theorems.

Theorem	Expression
Linearity	$\Im\{Ag(x,y) + Bh(x,y)\} = A\Im\{g(x,y)\} + B\Im\{h(x,y)\}$
Similarity	$\Im\left\{g\left(\dfrac{x}{a}, \dfrac{y}{b}\right)\right\} = \|ab\|G(af_X, bf_Y)$
Shift	$\Im\{g(x-a, y-b)\} = G(f_X, f_Y)\exp\left[-j2\pi(f_X a + f_Y b)\right]$
Parseval's (Rayleigh's)	$\iint \|g(x,y)\|^2 dxdy = \iint \|G(f_X, f_Y)\|^2 df_X df_Y$
Convolution	$\Im\left\{\iint g(\xi,\eta)h(x-\xi, y-\eta)d\xi d\eta\right\} = G(f_X, f_Y)H(f_X, f_Y)$
Autocorrelation	$\Im\left\{\iint g(\xi,\eta)g^*(\xi-x, \eta-y)d\xi d\eta\right\} = \|G(f_X, f_Y)\|^2$ $\Im\{\|g(x,y)\|^2\} = \iint G(\xi,\eta)G^*(\xi-f_X, \eta-f_Y)d\xi d\eta$
Cross-correlation	$\Im\left\{\iint g(\xi,\eta)h^*(\xi-x, \eta-y)d\xi d\eta\right\} = G(f_X, f_Y)H^*(f_X, f_Y)$ $\Im\{g(x,y)h^*(x,y)\} = \iint G(\xi,\eta)H^*(\xi-f_X, \eta-f_Y)d\xi d\eta$
Fourier integral	$\Im\Im^{-1}\{g(x,y)\} = \Im^{-1}\Im\{g(x,y)\} = g(x,y)$
Successive transform	$\Im\Im\{g(x,y)\} = g(-x,-y)$
Central ordinate	$\Im\{g(x,y)\}\big\|_{\substack{f_X=0 \\ f_Y=0}} = G(0,0) = \iint g(x,y)dxdy$ $\Im^{-1}\{G(f_X, f_Y)\}\big\|_{\substack{x=0 \\ y=0}} = g(0,0) = \iint G(f_X, f_Y)df_X df_Y$

Note: A, B, a, and b are scalar constants

1.5 Basic Functions and Transforms

Several basic functions, or combinations thereof, are used to describe various physical or analytic structures encountered in optics, such as a circle function to describe a circular aperture. Thus, these functions and their Fourier transform pairs are of considerable utility. The definitions in Table 1.2 are adopted.

Functions of one variable are illustrated in Fig. 1.1. These can be combined as products to represent separable 2D functions. The circle function is a symmetric 2D function where a single radial variable $r = (x^2+y^2)^{1/2}$ is often used. A shorthand name is not defined for the Gaussian, but this function appears often. The form we use is convenient for Fourier analysis. The circle and a 2D Gaussian function are plotted in Fig. 1.2 for illustration.

Table 1.2 Basic functions.

Function	Definition
Rectangle	$\mathrm{rect}(x) = \begin{cases} 1, & |x| < \frac{1}{2} \\ \frac{1}{2}, & |x| = \frac{1}{2} \\ 0, & \text{otherwise} \end{cases}$
Sinc	$\mathrm{sinc}(x) = \dfrac{\sin(\pi x)}{\pi x}$
Triangle	$\Lambda(x) = \begin{cases} 1 - |x|, & |x| \leq 1, \\ 0, & \text{otherwise.} \end{cases}$
Comb	$\mathrm{comb}(x) = \sum_{n=-\infty}^{\infty} \delta(x-n)$
Gaussian	$\exp(-\pi x^2)$
Circle	$\mathrm{circ}\left(\sqrt{x^2+y^2}\right) = \begin{cases} 1, & \sqrt{x^2+y^2} < 1 \\ \frac{1}{2}, & \sqrt{x^2+y^2} = 1 \\ 0, & \text{otherwise.} \end{cases}$

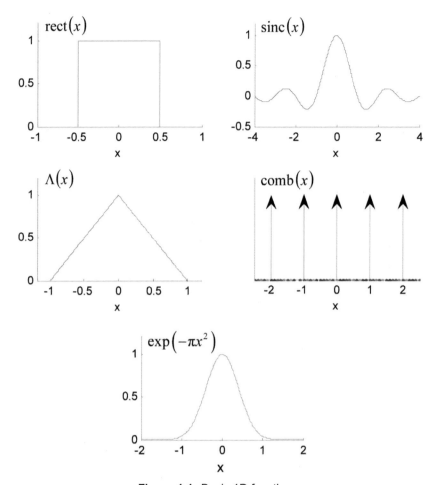

Figure 1.1 Basic 1D functions.

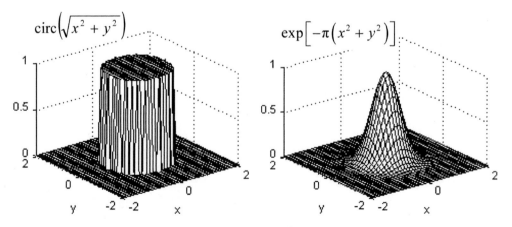

Figure 1.2 Examples of 2D functions.

If optical structures and apertures are modeled with basic functions, then corresponding Fourier transforms can aid in finding diffraction solutions or image results. The basic functions and their Fourier transforms are presented in Table 1.3. J_1 is a Bessel function of the first kind, order 1, and appears in the transform of the circle function. The transform of the circle is illustrated in Fig. 1.3. In Table 1.3, the last row gives a pair of "chirp" functions that will become quite familiar in the following chapters.

1.6 Linear and Space-Invariant Systems

The power of Fourier methods to analyze the response of a physical system to an input is significantly enhanced if the system can be modeled as linear and shift- (or space-) invariant. There are many aspects of optical systems that can be modeled in this way. In general, the operation of a system on a two-variable input

Table 1.3 Basic functions and their transforms.

Function	Transform
$\text{rect}\left(\dfrac{x}{a}\right)$	$\|a\|\,\text{sinc}(af_X)$
$\text{sinc}\left(\dfrac{x}{a}\right)$	$\|a\|\,\text{rect}(af_X)$
$\Lambda\left(\dfrac{x}{a}\right)$	$\|a\|\,\text{sinc}^2(af_X)$
$\text{comb}\left(\dfrac{x}{a}\right)$	$\|a\|\,\text{comb}(af_X)$
$\exp\left(-\pi\dfrac{x^2}{a^2}\right)$	$\|a\|\exp\left(-\pi a^2 f_X^2\right)$
$\text{circ}\left(\dfrac{\sqrt{x^2+y^2}}{a}\right)$	$a^2\,\dfrac{J_1\left(2\pi a\sqrt{f_X^2+f_Y^2}\right)}{a\sqrt{f_X^2+f_Y^2}}$
$\exp\left[-\pi\left(\dfrac{x^2}{a^2}+\dfrac{y^2}{b^2}\right)\right]$	$\|ab\|\exp\left[-\pi\left(a^2 f_X^2+b^2 f_Y^2\right)\right]$
$\exp\left[j\pi\left(\dfrac{x^2}{a^2}+\dfrac{y^2}{b^2}\right)\right]$	$j\|ab\|\exp\left[-j\pi\left(a^2 f_X^2+b^2 f_Y^2\right)\right]$

Note: J_1 is a Bessel function of the first kind, order 1.

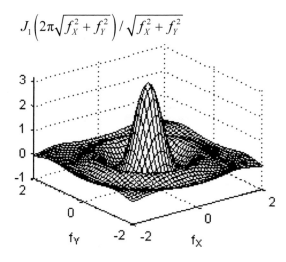

Figure 1.3 Circle function transform; peak value at $f_X = 0, f_Y = 0$ is π.

function g_1 to produce an output function g_2 can be described by

$$g_2(x_2, y_2) = S\{g_1(x_1, y_1)\}, \tag{1.5}$$

where S indicates the operation performed by the system. The "test" for *linearity* is the following:

$$S\{Ag_A(x_1, y_1) + Bg_B(x_1, y_1)\} = AS\{g_A(x_1, y_1)\} + BS\{g_B(x_1, y_1)\}, \tag{1.6}$$

where A and B are scalar constants. For a sum of input functions—for example, g_A and g_B in Eq. (1.6) with constant multipliers—the output is a sum of the individual responses. If the input can be "decomposed" into a sum of "elementary functions," then the output of a linear system can be determined if the response to the elementary functions is known. Linearity leads to the following expression, known as a *superposition integral*:

$$g_2(x_2, y_2) = \int\int_{-\infty}^{\infty} g_1(\xi, \eta) h(x_2, y_2; \xi, \eta) d\xi d\eta. \tag{1.7}$$

The function h is the *impulse response* of the system, and the integrals indicate that the output of the system is a superposition—or sum—of an infinite set of impulse responses that multiply the input function. The impulse response is modeled by

$$h(x_2, y_2; \xi, \eta) = S\{\delta(x_1 - \xi, y_1 - \eta)\}. \tag{1.8}$$

A linear system is completely characterized by its responses to impulse functions, but to use this property in practice the responses must be known for all locations in the input plane (x_1, y_1).

Linearity represents one level of simplification. Further simplification is afforded by the property of *space invariance*, where in its most basic form we write

$$g_2(x_2 - \xi, y_2 - \eta) = S\{g_1(x_1 - \xi, y_1 - \eta)\}, \tag{1.9}$$

and, therefore, the impulse response simplifies to

$$h(x_2, y_2; \xi, \eta) = h(x_2 - \xi, y_2 - \eta). \tag{1.10}$$

This impulse response does not depend on the absolute position in the input plane or the output plane. It only depends on the relative separation of the input and output points as if they were to appear in a common x–y plane. An interpretation of this situation is that an impulse anywhere in the input plane creates a corresponding response in the output plane that changes position with the input but always has the same relative form. The superposition integral now becomes a *convolution integral*:

$$g_2(x_2, y_2) = \int\int_{-\infty}^{\infty} g_1(\xi, \eta) h(x_2 - \xi, y_2 - \eta) d\xi d\eta. \tag{1.11}$$

In shorthand notation with the convolution operator $\otimes$, Eq. (1.11) is written as

$$g_2(x, y) = g_1(x, y) \otimes h(x, y), \tag{1.12}$$

where the subscripts on the x and y variables are no longer necessary. Taking the Fourier transform of each function in Eq. (1.12) and applying the convolution theorem yields

$$G_2(f_X, f_Y) = G_1(f_X, f_Y) H(f_X, f_Y), \tag{1.13}$$

where $H(f_X, f_Y)$ is the Fourier transform of the impulse response $h(x, y)$ and is known as the *transfer function*. Two valuable features of *linear space-invariant systems* are apparent from Eq. (1.13):

1. Rather than directly tackling the convolution integral, a more computationally appealing route can be taken of transforming the input to the Fourier domain, multiplying by the transfer function, and inverse transforming to find the result.
2. The transfer function model is analogous to frequency-filtering operations that are found in electric circuit theory, digital signal

processing, and many other disciplines involving signal analysis. Insights from these areas can often be applied to Fourier methods associated with optical systems.

1.7 Exercises

1.1 Sketch the following functions:

(a) $\operatorname{rect}\left(\dfrac{x}{2}\right)$,

(b) $\operatorname{rect}(x-2)$,

(c) $\Lambda\left(\dfrac{x+2}{2}\right)$,

(d) $\exp(-3\pi x^2)$,

(e) $\left[\operatorname{comb}\left(\dfrac{x}{4}\right) * \Lambda(x)\right]\operatorname{rect}\left(\dfrac{x}{12}\right)$,

(f) $\operatorname{circ}\left(\sqrt{(x-2)^2+y^2}\right) + \operatorname{circ}\left(\sqrt{(x+2)^2+y^2}\right)$.

1.2 Using known transform pairs and theorems, find the Fourier transforms of the following:

(a) $\operatorname{rect}\left(\dfrac{x}{2w}\right)\operatorname{rect}\left(\dfrac{y}{2w}\right)$,

(b) $\operatorname{rect}\left(\dfrac{x-x_0}{2w}\right)\operatorname{rect}\left(\dfrac{y}{2w}\right)$,

(c) $\exp\left(-\dfrac{x^2+y^2}{w^2}\right)$,

(d) $\operatorname{circ}\left(\dfrac{\sqrt{x^2+y^2}}{w_1}\right) - \operatorname{circ}\left(\dfrac{\sqrt{x^2+y^2}}{w_2}\right)$,

(e) $\operatorname{circ}\left(\dfrac{\sqrt{(x-d/2)^2+y^2}}{w}\right) + \operatorname{circ}\left(\dfrac{\sqrt{(x+d/2)^2+y^2}}{w}\right)$.

1.3 Perform the following convolutions by applying the convolution theorem:

(a) $\quad \text{rect}\left(\dfrac{x}{2w}\right)\text{rect}\left(\dfrac{y}{2w}\right) \otimes \text{rect}\left(\dfrac{x}{2w}\right)\text{rect}\left(\dfrac{y}{2w}\right),$

(b) $\quad \exp\left(-\pi\dfrac{x^2+y^2}{3^2}\right) \otimes \exp\left(-\pi\dfrac{x^2+y^2}{4^2}\right),$

(c) $\quad \text{sinc}\left(\dfrac{x}{2}\right)\text{sinc}(y) \otimes \text{sinc}\left(\dfrac{x}{4}\right)\text{sinc}(y).$

1.4 Find the autocorrelations of the following:

(a) $\quad \text{rect}\left(\dfrac{x}{2w}\right)\text{rect}\left(\dfrac{y}{2w}\right),$

(b) $\quad \exp\left(-\pi\dfrac{x^2+y^2}{w^2}\right).$

1.5 Apply the central ordinate theorem to find $\iint g(x,y)\,dxdy$ for the following, and compare the results with simple area calculations:

(a) $\quad g(x,y) = \text{rect}\left(\dfrac{x}{2w}\right)\text{rect}\left(\dfrac{y}{4w}\right),$

(b) $\quad g(x,y) = \text{circ}\left(\dfrac{\sqrt{x^2+y^2}}{3}\right).$

1.6 Demonstrate whether the following operations are linear and/or space-invariant, where A, B, a, and b are scalar constants:

(a) $\quad S\{g(x,y)\} = Ag(x,y),$

(b) $\quad S\{g(x,y)\} = Ag(x,y)+B,$

(c) $\quad S\{g(x,y)\} = A[g(x,y)]^2,$

(d) $\quad S\{g(x,y)\} = xg(x,y),$

(e) $\quad \text{Ave}\{g(x,y)\} \equiv \dfrac{1}{ab}\int_{x-\frac{a}{2}}^{x+\frac{a}{2}}\int_{y-\frac{b}{2}}^{y+\frac{b}{2}} g(\xi,\eta)\,d\xi d\eta.$

1.8 References

1. W. T. Rhodes, "History and evolution of the teaching of Fourier optics," *Proc. SPIE,* **3572,** 50–56 (1999). [doi:10.1117/12.358418].
2. J. W. Goodman, *Introduction to Fourier Optics*, 3rd Ed., Roberts & Company, Greenwood Village, CO (2005).
3. R. Barakat, "The calculation of integrals encountered in optical diffraction theory," in *The Computer in Optical Research, Topics in Applied Physics,* Vol. **41,** B. R. Frieden (ed.), Springer, Berlin (1980).
4. J. D. Gaskill, *Linear Systems, Fourier Transforms, and Optics*, Wiley-Interscience, New York (1978).
5. O. K. Ersoy, *Diffraction, Fourier Optics and Imaging*, Wiley-Interscience, New York (2006).
6. C. A. Bennett, *Principles of Physical Optics*, Wiley, Hoboken, NJ (2008).
7. E. G. Steward, *Fourier Optics: An Introduction*, 2nd Ed., Dover, New York (2004).
8. W. J. Smith, *Modern Optical Engineering*, 4th Ed., McGraw-Hill Professional, New York (2007).
9. J. M. Geary, *Introduction to Lens Design with Practical ZEMAX® Examples*, Willmann-Bell, Richmond, VA (2002).

Chapter 2
Sampled Functions and the Discrete Fourier Transform

When implementing Fourier optics simulations on the computer it is necessary to represent functions by discrete arrays of sampled values and apply transform and processing methods designed for these discrete signals. To come as close as possible to simulating continuous space, it would be great to model the physical elements with a gazillion samples. However, computer memory and execution time limitations won't allow this. Thus, devising practical Fourier optics simulations becomes an act of balancing acceptable sampling artifacts and available computer resources. This chapter begins to address this matter with discussions of the sampling of continuous functions, the Shannon–Nyquist sampling theorem, and the concept of effective bandwidth. The remainder of the chapter concerns the discrete Fourier transform (DFT), the workhorse tool for computational Fourier analysis. We look at its relationship with the analytic transform, describe implementation details, and discuss how DFT results differ from the analog world.

2.1 Sampling and the Shannon–Nyquist Sampling Theorem

Consider the two-dimensional (2D) analytic function $g(x,y)$ and suppose it is sampled in a uniform manner (Fig. 2.1) in the x and y directions, which is indicated by

$$g(x, y) \rightarrow g(m\Delta x, n\Delta y), \tag{2.1}$$

where the *sample interval* is Δx in the x direction and Δy in the y direction, and m and n are integer-valued indices of the samples. The respective *sample rates* are $1/\Delta x$ and $1/\Delta y$. In practice, the sampled space is finite and, assuming it is composed of $M \times N$ samples in the x and y directions, respectively, m and n are often defined with the following values:

$$m = -\frac{M}{2}, \ldots, \frac{M}{2} - 1, \quad n = -\frac{N}{2}, \ldots, \frac{N}{2} - 1. \tag{2.2}$$

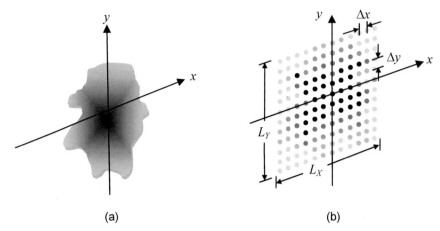

Figure 2.1 Two-dimensional function: (a) analytic and (b) sampled versions.

This is a standard index arrangement where M and N are assumed to be even. Even numbers of samples are used in this book for reasons associated with discrete Fourier transform efficiency and sample arrangement (see Section 2.4).

A finite physical area (e.g., units of m^2) is spanned by the sampled space, and this is given by $L_X \times L_Y$, where L_X is the length along the x side of the sampled space and L_Y is the length along the y side (Fig. 2.1). L_X and L_Y are referred to as the *side lengths*. They represent physical distances and are related to the sampling parameters by

$$L_X = M\Delta x, \quad L_Y = N\Delta y. \tag{2.3}$$

An obvious sampling concern is whether all the *significant values* of $g(x,y)$ "fit" within the physical area defined by $L_X \times L_Y$. The *support* of $g(x,y)$ refers to the span of the significant values. This concept is illustrated in Fig. 2.2(a) for one axis. If D_X is the support in the x direction and D_Y is the support in the y direction, then for the significant values of $g(x,y)$ to be contained within the array requires

$$D_X < L_X, \quad D_Y < L_Y. \tag{2.4}$$

Another concern is whether the sample intervals are small enough to preserve the features of $g(x,y)$. For functions that are *bandlimited*, where the spectral content of the signal is limited to a finite range of frequencies, a continuous function can be recovered exactly from the samples if the sample interval is smaller than a specific value. The Shannon–Nyquist sampling theorem, extended to two dimensions, states this requirement as[1]

$$\Delta x < \frac{1}{2B_X}, \quad \Delta y < \frac{1}{2B_Y}, \tag{2.5}$$

Sampled Functions and the Discrete Fourier Transform

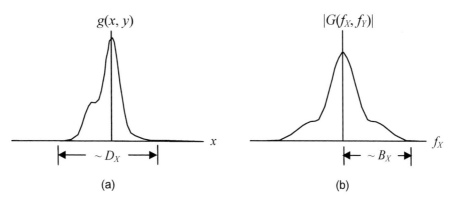

Figure 2.2 Illustration of the (a) support D_X and (b) bandwidth B_X along the x axis of $g(x, y)$. Bandwidth is commonly defined as a half-width measure and is illustrated here with a profile of $|G(f_X, f_Y)|$, the Fourier transform magnitude of $g(x, y)$.

where B_X is the bandwidth of the spectrum of the continuous function along the x direction and B_Y is the bandwidth along the y direction. Bandwidth is illustrated in Fig. 2.2(b). Violating Eq. (2.5) results in *aliasing*, in which undersampled high-frequency components in the signal are interpreted erroneously as low-frequency content. This issue is considered further in Section 2.5. A related parameter is the *Nyquist frequency* given by

$$f_{NX} = \frac{1}{2\Delta x}, \quad f_{NY} = \frac{1}{2\Delta y}, \quad (2.6)$$

which is half the sample rate and corresponds to the maximum spatial frequency that can be adequately represented given the interval Δx or Δy.

2.2 Effective Bandwidth

Practical functions such as those defined in Chapter 1 are not bandlimited. In fact, any function with finite support, like the rectangle or circle functions, cannot be bandlimited. Often these functions have an *effective bandwidth* that encompasses the most significant frequency values. Even though the criteria posed by the Shannon–Nyquist theorem may not be completely satisfied, a small enough sample interval can be found to provide an acceptable representation of the analytic function where the effects of aliasing are small.

For example, consider a 2D square signal with half-width of w:

$$f(x, y) = \text{rect}\left(\frac{x}{2w}\right)\text{rect}\left(\frac{y}{2w}\right). \quad (2.7)$$

The analytic Fourier transform yields the spectrum

$$F(f_X, f_Y) = 4w^2 \operatorname{sinc}(2wf_X) \operatorname{sinc}(2wf_Y). \tag{2.8}$$

One approach to defining the effective bandwidth is to find the spectral width (radius) that contains a high percentage of the total power in the spectrum. Applying Parseval's theorem, the total spectral power of Eq. (2.8) is

$$\begin{aligned} P_T &= \int\int_{-\infty}^{\infty} (4w^2)^2 \operatorname{sinc}^2(2wf_X) \operatorname{sinc}^2(2wf_Y) df_X df_Y \\ &= \int\int_{-\infty}^{\infty} \operatorname{rect}^2\left(\frac{x}{2w}\right) \operatorname{rect}^2\left(\frac{y}{2w}\right) dx dy = 4w^2 \end{aligned} \tag{2.9}$$

A practical criterion for the effective bandwidth B is to include 98% of the total spectral power. Converting to polar coordinates to allow a radial bandwidth value B to be considered leads to

$$\frac{1}{P_T} \int_0^{2\pi}\int_0^B (4w^2)^2 \operatorname{sinc}^2[2w(\rho\cos\theta)] \operatorname{sinc}^2[2w(\rho\sin\theta)] \rho\, d\rho\, d\theta = 0.98, \tag{2.10}$$

where $f_X = \rho \cos\theta$ and $f_Y = \rho \sin\theta$. The integrals on the left side can be evaluated numerically for various values of B until Eq. (2.10) is satisfied. With this approach the effective bandwidth is found to be

$$B \approx \frac{5}{w}. \tag{2.11}$$

Figure 2.3 illustrates the portion of the spectrum that encompasses 98% of the spectral power. Substituting Eq. (2.11) into Eq. (2.5) for the bandwidth gives

$$\Delta x \le \frac{w}{10}, \tag{2.12}$$

which says at least 10 samples across the half-width of the rect function (20 across the full width) are required to retain the effective bandwidth indicated in Eq. (2.11). It is important to realize that the part of the analytic spectrum that lies beyond the Nyquist frequency does not simply disappear. Even though small in power, it can introduce noticeable aliased frequency content that is erroneous.

Table 2.1 shows effective bandwidth values for square, circle, and Gaussian functions computed in the same way. A larger effective bandwidth can be used if there is a need to include more of the spectral power.

Sampled Functions and the Discrete Fourier Transform 17

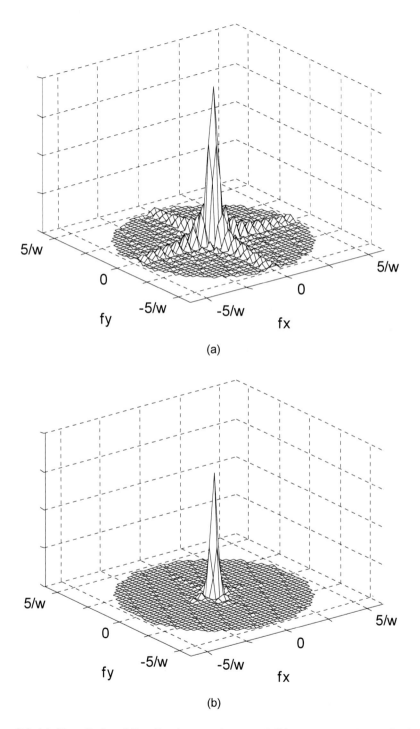

Figure 2.3 (a) Magnitude of the Fourier spectrum and (b) power spectrum of $g(x,y) =$ rect(x/w) rect(y/w), comprising 98% of the total spectral power.

2.3 Discrete Fourier Transform from the Continuous Transform

The DFT, usually in the form of its highly efficient offspring—the fast Fourier transform (FFT)—is a fundamental tool for modeling Fourier optics problems on the computer. The objective for this section is to develop the DFT, a discrete implementation of the continuous Fourier transform. The derivation is helpful for understanding the scaling of the spatial sample coordinates and frequency coordinates, as well as the constants multiplying a discrete transform result. This scaling is an important part of modeling a physical optics problem. Only the aspects of the DFT and FFT that are critical to the simulation approaches covered in this book are highlighted, so there are many more details to be discovered (or rediscovered) in other resources.[2-4]

The analytic Fourier transform of a function g of two variables x and y is repeated here for reference

$$G(f_X, f_Y) = \int_{-\infty}^{\infty}\int g(x,y)\exp\left[-j2\pi(f_X x + f_Y y)\right]dxdy. \quad (2.13)$$

First, assume $g(x,y)$ is sampled as indicated in Eqs. (2.1) and (2.2). To simplify some of the notation, the following substitution can be used where the actual sample intervals are not explicitly shown:

$$g(m\Delta x, n\Delta y) \rightarrow \tilde{g}(m,n). \quad (2.14)$$

Next, the integrals in Eq. (2.13) can be approximated using a Riemann sum:

$$\int_{-\infty}^{\infty}\int_{-\infty}^{\infty}\ldots dxdy \rightarrow \sum_{n=-N/2}^{N/2-1}\sum_{m=-M/2}^{M/2-1}\ldots \Delta x \Delta y. \quad (2.15)$$

Table 2.1 Effective bandwidth for 98% power.

Function	Effective Bandwidth B
$\text{rect}\left(\dfrac{x}{2w}\right)\text{rect}\left(\dfrac{y}{2w}\right)$	$\dfrac{5}{w}$
$\text{circ}\left(\dfrac{\sqrt{x^2+y^2}}{w}\right)$	$\dfrac{5}{w}$
$\exp\left[-\pi\left(\dfrac{x^2+y^2}{w^2}\right)\right]$	$\dfrac{0.79}{w}$

Because the DFT operation is performed generically on a discrete array of values without specific sample interval information, the multipliers $\Delta x \Delta y$ in Eq. (2.15) are not included in the DFT definition. However, these multipliers need to be applied subsequently to the DFT operation for appropriate scaling of a physical problem.

The convention for the frequency domain is to divide this continuous "space," indicated by f_X and f_Y, into M and N evenly spaced coordinate values. This involves the following substitutions:

$$f_X \to \frac{p}{M\Delta x}, \quad \text{where} \quad p = -\frac{M}{2}, \ldots, \frac{M}{2} - 1;$$

$$f_Y \to \frac{q}{N\Delta y}, \quad \text{where} \quad q = -\frac{N}{2}, \ldots, \frac{N}{2} - 1; \quad (2.16)$$

where p and q are integers that index multiples of the *frequency sample intervals*

$$\Delta f_X = \frac{1}{M\Delta x} = \frac{1}{L_X}, \quad \text{and} \quad \Delta f_Y = \frac{1}{N\Delta y} = \frac{1}{L_Y}. \quad (2.17)$$

In fact, p and q take on the same values as m and n, respectively, since the spatial and frequency arrays have the same number of elements. Note that the maximum absolute values of the frequency coordinates in Eq. (2.16) are the Nyquist frequencies $1/(2\Delta x) = f_{NX}$ and $1/(2\Delta y) = f_{NY}$. Incorporating Eq. (2.16) into the complex exponential kernel of Eq. (2.13) yields

$$\exp[-j2\pi(f_X x + f_Y y)] \to \exp\left[-j2\pi\left(\frac{p}{M\Delta x} m\Delta x + \frac{q}{N\Delta y} n\Delta y\right)\right]$$

$$= \exp\left[-j2\pi\left(\frac{pm}{M} + \frac{qn}{N}\right)\right]. \quad (2.18)$$

Finally, substituting Eqs. (2.14)–(2.18) into Eq. (2.13), we arrive at the following form of the DFT:

$$\tilde{G}(p,q) = \sum_{m=-M/2}^{M/2-1} \sum_{n=-N/2}^{N/2-1} \tilde{g}(m,n) \exp\left[-j2\pi\left(\frac{pm}{M} + \frac{qn}{N}\right)\right], \quad (2.19)$$

where $\tilde{G}(p,q)$ represents the DFT of $\tilde{g}(m,n)$. The inverse discrete Fourier transform (DFT^{-1}) is derived in a similar way and is written as

$$\tilde{g}(m,n) = \frac{1}{MN} \sum_{p=-M/2}^{M/2-1} \sum_{q=-N/2}^{N/2-1} \tilde{G}(p,q) \exp\left[j2\pi\left(\frac{pm}{M} + \frac{qn}{N}\right)\right]. \quad (2.20)$$

The appearance of the $1/MN$ multiplier in Eq. (2.20) requires some discussion. The factor is equal to the product $\Delta x \Delta y \Delta f_X \Delta f_Y$ [see Eq. (2.17)], which comes about when numerically evaluating a forward, and then an inverse Fourier integral in succession. This factor allows the DFT followed by the DFT^{-1} to return the original function values, which is consistent with the Fourier integral theorem. The application of $1/MN$ varies in different software tools. For example, MATLAB implements the transform and inverse transform based on the definitions in Eqs. (2.19) and (2.20), but some applications apply a $(MN)^{-\frac{1}{2}}$ factor to both the forward and inverse transforms. In some modeling situations we will need to account for this factor.

The forward and inverse DFTs are not usually accomplished with a direct execution of Eqs. (2.19) and (2.20), but rather they are accomplished with the computationally efficient FFT and FFT^{-1} algorithms. These algorithms implement a scheme that is not of specific importance here other than to say that the result is consistent with Eqs. (2.19) and (2.20). FFT algorithms are most efficient when M and N are a power of 2, although computation times can be nearly as fast for other values. A practical issue for FFT implementation concerns the arrangement and indexing of data values in an array. This issue is now discussed below.

2.4 Coordinates, Indexing, Centering, and Shifting

Uniform sampling and square grids, where $\Delta y = \Delta x$, $N = M$, and $L_X = L_Y = L$, are often used in practice. This will be the case for all the examples presented in this book; so, to simplify the presentation often only one set of variables is discussed.

Considering Eqs. (2.1) and (2.2), the *coordinates* of the samples along one dimension can be described as

$$x \rightarrow \left[-\frac{L}{2} \; : \; \Delta x \; : \; \frac{L}{2} - \Delta x\right], \quad (2.21)$$

where the above notation is borrowed from MATLAB and indicates that the coordinates range from $-L/2$ to $L/2 - \Delta x$ in steps of Δx. The y-axis coordinates are defined similarly. Assuming a FFT relationship between the spatial and spectral domains, then from Eqs. (2.16) and (2.17) the following is derived:

$$f_X \rightarrow \left[-\frac{1}{2\Delta x} \; : \; \frac{1}{L} \; : \; \frac{1}{2\Delta x} - \frac{1}{L}\right], \quad (2.22)$$

which indicates that the spatial frequency coordinates range from $-1/(2\Delta x)$ to $1/(2\Delta x)-1/L$ in steps of $\Delta f_X = 1/L$. Again, the f_Y coordinates are similar.

The integer index variables m and n, as well as p and q introduced in Eqs. (2.2) and (2.16) span negative as well as positive values. However, software applications use positive integer values for vector or array (matrix) indexing. In the case of MATLAB, indexing for a one-dimensional (1D) vector begins at (1). For display purposes it is convenient to "center" the function of interest in the vector, which means the zero coordinate will correspond to the ($M/2+1$) index. However, the convention for the 1D FFT algorithms is that the data value placed in the first index position corresponds to the zero coordinate. Thus, a "shift" of the centered vector values is needed before an FFT operation.

Figure 2.4 illustrates arrangements of values and indices of a 1D sampled rect function and its spectrum. The arrangement in Figs. 2.4(a) and (b) is consistent with the analytic development in Section 2.1, where indices span negative and positive values. Figures 2.4(c) and (d) illustrate the *centered* arrangement in the computer with positive indices, and Figs. 2.4(e) and (f) show the *shifted* arrangement that is necessary prior to an FFT or FFT^{-1}. Conversion between the centered and shifted arrangements can be done easily with the MATLAB command `fftshift`. Reversing the shift to get back to the centered arrangement is done with `ifftshift`.

For a 2D array, MATLAB indexing begins at (1,1). A centered function has the zero-coordinate $[x,y] = [0,0]$ value located at index ($N/2+1$, $M/2+1$). Prior to the 2D FFT, a shift is needed to place the zero-coordinate value at index (1,1). Figure 2.5 illustrates the centered and shifted arrangements for 2D arrays. In Fig. 2.5, row order is top-to-bottom, column order is left-to-right. The required shift is actually a swapping of array quadrants. Again, `fftshift` and `ifftshift` can be applied to move between the centered and shifted arrangements.

A confusing detail for 2D array data is that MATLAB uses a row–column indexing scheme, where (i, j) indicates the ith row and the jth column. This is, in a sense, a reverse of standard Cartesian coordinate notation, where x (horizontal axis or "column") is listed first and y (vertical axis or "row") is listed second. Thus, the j-indices correspond to x-coordinates and i-indices correspond to y-coordinates. This explains the (j, i) listing ($N/2+1$, $M/2+1$) paired with $[x,y]$ values in Fig. 2.5. Fortunately, MATLAB's vector operation notation is developed for the Cartesian coordinate system; so, this issue is mostly transparent as far as programming is concerned. It is only when the actual integer index values are used in codes that this array arrangement becomes an issue.

2.5 Periodic Extension

Roughly speaking, all of the Fourier transform theorems listed in Table 1.1 can be applied in the discrete domain. For example, a convolution can be performed by computing the FFTs of two discrete functions, multiplying the results (point-wise) and computing the FFT^{-1}. However, discrete transform results differ from analytic results in a way that is characterized by a concept known as periodic

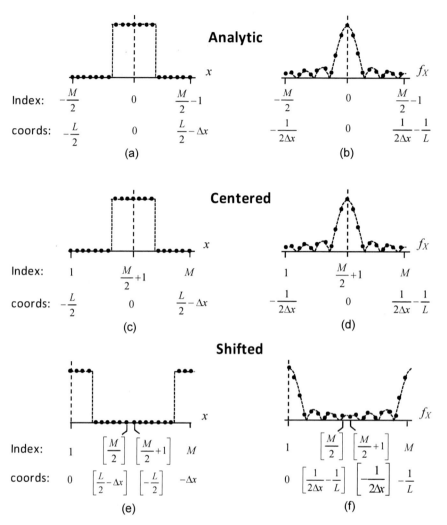

Figure 2.4 Sampling arrangements for 1D spatial (left column) and frequency (right column) vectors: (a) and (b) analytic; (c) and (d) centered; and (e) and (f) shifted for FFT operations.

extension. Here, we provide a short discussion of this property and an illustration. The topic is covered in more depth in other resources, such as the work by Brigham.[4] Consider the 1D, analytic function $f(x)$ and a sampled version given by

$$f_s(x) = \left[f(x) \cdot \text{comb}\left(\frac{x}{\Delta x}\right) \right] \text{rect}\left(\frac{x}{L}\right). \quad (2.23)$$

Sampled Functions and the Discrete Fourier Transform

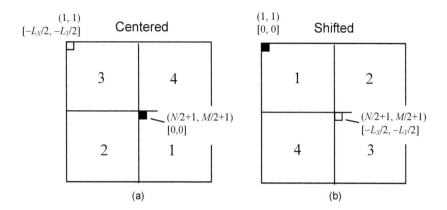

Figure 2.5 Sampling arrangements for 2D spatial array: (a) centered and (b) shifted for the 2D FFT.

The comb function "picks out" the sample values at intervals of Δx and the rect function sets the overall sampled space as L. Taking the analytic Fourier transform of Eq. (2.23) gives

$$F_s(f_X) = \left[F(f_X) * \Delta x \operatorname{comb}(\Delta x f_X) \right] * L \operatorname{sinc}(L f_X), \qquad (2.24)$$

where $*$ indicates a 1D convolution. This result is a continuous function where the analytic spectrum $F(f_X)$ is repeated at intervals of $1/\Delta x$. The sinc convolution is a "smoothing" process. However, the FFT operation actually produces a sampled result that can be modeled by altering Eq. (2.24) with a sampling term

$$F_P(f_X) = \left\{ \left[F(f_X) * \Delta x \operatorname{comb}(\Delta x f_X) \right] * L \operatorname{sinc}(L f_X) \right\} \cdot L \operatorname{comb}(L f_X). \qquad (2.25)$$

The new comb term sets the sample spacing in the frequency domain to be $1/L$. This is consistent with Eq. (2.17). By inverse transforming Eq. (2.25), we find the function that corresponds to the spectrum $F_P(f_X)$:

$$f_P(x) = \left\{ \left[f(x) \cdot \operatorname{comb}\left(\frac{x}{\Delta x}\right) \right] \operatorname{rect}\left(\frac{x}{L}\right) \right\} * \operatorname{comb}\left(\frac{x}{L}\right). \qquad (2.26)$$

So, the periodic extension concept can be described as follows: although we start with a sampled version of $f(x)$ in the spatial domain, when the FFT is performed, it is as if we started with the periodic function $f_P(x)$ and produced the periodic spectrum $F_P(f_X)$.

To illustrate, an analytic rectangle function is shown in Fig. 2.6 (solid line) along with a sampled version (dots). The sampled version is contained in a vector of length 20, where $L = 20$ and $\Delta x = 1$. The periodic form of the function, which extends (virtually) beyond the original span of the sample vector, is also indicated (dashed line). Figure 2.6(b) shows the magnitude of the analytic spectrum of the rectangle (solid), the FFT result (dots), and the periodic spectrum (dashed). Figure 2.6(b) illustrates the FFT samples follow the periodic spectrum. The most obvious difference between the analytic and sample spectra in this case is slightly larger sample values in the magnitude lobes at higher frequencies. This effect results from aliasing of undersampled frequencies in the rectangle spectrum. The periodic extension concept is an instructive way to define this effect. In practice, by sampling a function to preserve the effective bandwidth—for example, the 98% power level—only a small amount of aliasing is introduced.

2.6 Periodic Convolution

Another issue discussed in the context of periodic extension is *periodic convolution*. If the FFT is applied to perform the convolution $f(x)*h(x)$, the result is actually a scaled form of $f_P(x)*h_P(x)$, where the periodic form $f_P(x)$ is

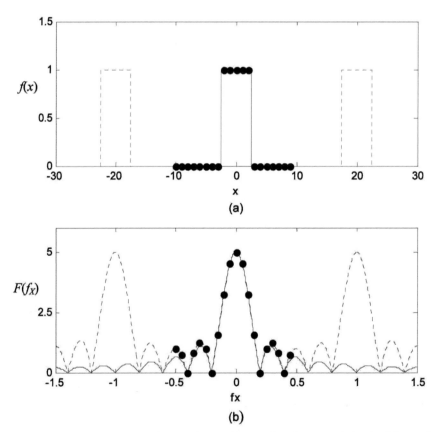

Figure 2.6 Periodic extension illustration: (a) rect function—analytic (solid), periodic extension (dash) and sampled (dot); and (b) rect spectral magnitude—analytic (solid), periodic extension (dash) and FFT samples (dot).

convolved with the periodic form $h_P(x)$. This situation is illustrated in Fig. 2.7 for rectangle and right triangle functions. Sampled versions of both functions are contained in vectors of 20 samples [Figs. 2.7(a) and (b)]. Conceptually, a convolution requires one function to be reversed, or "flipped," relative to the axis. The reversed function is translated across the second function, and the overlap area is recorded. Figure 2.7(c) shows the right triangle at a particular position in translation across the rectangle. The overlap area for an analytic convolution is the small gray triangle. However, because of periodic extension, a strip on the left side of the triangle is also overlapping a rectangle copy. This area is included in the discrete convolution result. As the right-triangle continues translating to the right, the erroneous area drops away, but the edges of the convolution result [dots in Fig. 2.7(d)] follow the periodic result.

In order for the periodic convolution to match the analytic convolution, the combined support of the two functions being convolved needs to be less than the array side length, or

$$D_1 + D_2 < L. \qquad (2.27)$$

For the example in Fig. 2.7, $D_1 = 9$ and $D_2 = 15$, but $L = 20$; so, Eq. (2.27) is violated, and the FFT-derived result does not match the analytic result.

2.7 Exercises

2.1 For a sample interval of $\Delta x = 10$ μm and side length $L = 5$ mm, what is the sample number M? What is the Nyquist frequency? What is the frequency sample interval? What is the range of coordinates in the spatial domain? What is the range of the coordinates in the frequency domain?

2.2 Consider the following:

(a) $\quad g(x,y) = \text{circ}\left(\dfrac{\sqrt{x^2 + y^2}}{w}\right), \quad w = 1$ mm,

(b) $\quad g(x,y) = \exp\left(-\dfrac{x^2 + y^2}{w^2}\right), \quad w = 1$ mm.

For each function determine the following: (1) the effective bandwidth; (2) the maximum sample interval Δx necessary to satisfy the sampling theorem given the effective bandwidth; and (3) assuming 256 samples (linear dimension), the maximum side length that can be modeled.

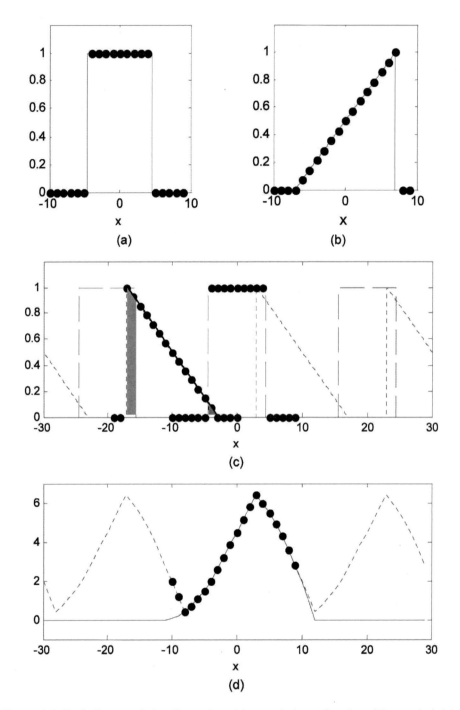

Figure 2.7 Periodic convolution illustration: (a) sampled rect function; (b) sampled right-triangle function; (c) shift and overlap illustration—periodic functions (dash) and sampled (dot) functions; and (d) convolution results—analytic (solid), periodic (dash) and FFT (dot).

2.3 What is the support of the following function along one dimension if the support is defined by where the function value drops to 1% of its peak?

$$g(x,y) = \exp\left[-\pi \frac{(x^2 + y^2)}{w^2}\right].$$

2.4 What is the bandwidth along one axis for the following?

(a) $\quad \mathrm{sinc}\left(\dfrac{x}{w}\right) \mathrm{sinc}\left(\dfrac{y}{w}\right),$

(b) $\quad \mathrm{sinc}^2\left(\dfrac{x}{w}\right) \mathrm{sinc}^2\left(\dfrac{y}{w}\right).$

2.5 Consider the following two functions:

$$g_1(x,y) = \Lambda\left(\frac{x}{d}\right)\Lambda\left(\frac{y}{2d}\right),$$

$$g_2(x,y) = \mathrm{circ}\left(\frac{\sqrt{x^2 + y^2}}{d}\right).$$

(a) What is the minimum side length required to accommodate a convolution of these two functions?

(b) What is the minimum side length required to accommodate the autocorrelation of $g_2(x, y)$?

2.8 References

1. J. W. Goodman, *Introduction to Fourier Optics*, 3rd Ed., Roberts & Company, Greenwood Village, CO (2005).
2. R. M. Gray and J. W. Goodman, *Fourier Transforms: An Introduction for Engineers*, Kluwer Academic, Boston (1995).
3. A. V. Oppenheim and R. W. Schafer, *Discrete-Time Signal Processing*, 3rd Ed., Prentice-Hall, Upper Saddle River, NJ (2009).
4. E. Brigham, *The Fast Fourier Transform and its Applications*, Prentice-Hall, Upper Saddle River, NJ (1988).

Chapter 3
MATLAB Programming of Functions, Vectors, Arrays, and Fourier Transforms

By following the examples presented in this chapter, you will gain some familiarity with MATLAB, learn how to implement a function, compute the discrete Fourier transform, and compare the result with analytic theory. For instructional purposes, most of the examples are one-dimensional (1D) problems, but two dimensions are introduced at the end of the chapter. MATLAB is a mathematics and graphics software application with its own interpreted language that is widely used for simulation and modeling in science and engineering disciplines. It is optimized for vector and matrix operations and, therefore, is a good tool for Fourier optics simulations, which generally involve at least two dimensions. The examples in this chapter and throughout the book use a basic set of MATLAB features, in part to keep the material at a tutorial level but also because the details of the programming steps are more obvious. As you become familiar with the software, you may find more convenient and efficient ways to implement the programming. MATLAB Version 7.1 is used in this book.

3.1 Defining Functions

Open MATLAB. The windows that are commonly displayed include the "Current Directory," "Command Window," and "Command History." These windows are often grouped together as part of the main window that comprises the MATLAB "Desktop" (Fig. 3.1). The Current Directory shows the folder in which your work will be stored and MATLAB-related files that are in that folder. Code can be entered in the Command Window where it is executed a line at a time as it is entered. Numerical output, such as the value of a variable, can also be displayed in the Command Window. The Command History shows a compact listing of the commands that have been entered in the Command Window. Another important window that can be called up is the "Editor Window," where

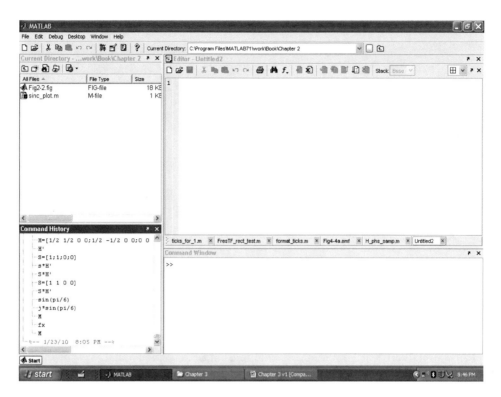

Figure 3.1 The MATLAB Version 7.1 desktop screen shows the Current Directory window (upper left), the Command History window (lower left), the Editor ("docked" in the upper right of the Desktop by clicking the down-curving arrow in the Editor window), and the Command Window (lower right).

code can be entered and saved as a file (an M-file with extension ".m"). Two different types of M-files can be created: a script or a function. A *script* is a collection of code lines that can be executed as a program. When a script is executed, numerical output shows up in the Command Window. *Function* M-files can be called from the Command Window or from a script to do a particular task and return a result. The examples presented here are described as being entered in the Editor Window, which allows your script to be saved in an M-file; but, the code can also be entered directly and executed in the Command Window.

The first thing to do is make a new folder for your work. On the toolbar above the Current Directory, click on the "New Folder" icon and a folder should appear with a cursor positioned to enter the folder name. Type in a name, such as "Fourier Optics," and click anywhere on the folder line or hit Enter. Then double click on this folder in the Current Directory window to open it as your current storage area.

The first programming step will be to create a rectangle (rect) function. Once created, this piece of code can be called from a script to generate a vector containing a sampled rectangle function. Call up the Editor Window by clicking on the New M-file icon (a document-like icon), which is on the far left of the

MATLAB main Desktop toolbar. A cursor next to the line number 1 should appear in the window. Type in the following:

```
1   function[out]=rect(x);
2   % rectangle function
3
4   out=abs(x)<=1/2;
5   end
```

To do this, simply type out each line, and hit the Enter key at the end of each line. MATLAB uses a simple text format, so special function keys are not required for entering the code. The "`%`" character indicates that the text that follows is a comment and not to be interpreted for execution. Some of the text in the display will be colored. Comment text is displayed in green. As always, documentation of your code is important and a few comments are included in the examples presented throughout this book; but, in general, they are kept to a minimum to keep the presentation concise. Comments can also be added on the same line following the code.

Blue-colored text indicates built-in MATLAB functions. The `function` command sets up this M-file to be a function named "`rect`" with an input vector "`x`" and an output vector "`out`." MATLAB is most proficient with vector and matrix operations, so the code tends to involve variables that are vectors and arrays as opposed to single parameters. The code in line 4 uses a vectorized logical test feature in MATLAB. The "`abs`" command takes the absolute value of each x element and "`<=`" applies the "less than or equal" test to each element. If the test is "true" (less than or equal to 1/2), then 1 is returned for that element. If the test is "false," a 0 is returned. The vector `out` has the same number of elements as x and will contain the sampled rectangle function. The semicolon at the end of a line suppresses the output in the Command Window as the code is executing.

In the `rect` code, the samples at the edges of the rectangle are not allowed to take on a value of 1/2 as in the analytic rectangle function definition (see Section 1.5). Doing so can be interpreted as a "slope" at the edge rather than a sharp transition, so we choose to work with a single point transition. The coding we use for `rect` has the characteristic that the full width of the function is always created with an odd number of samples. If a situation comes about where a rectangle with an even number of samples is requested, one more sample than expected will be returned. Some of the examples in the book use a width or radius value that is not a round number—this is simply to be consistent with the odd number of samples that will return from calling `rect`.

When you are done typing the code, click on "Save" (floppy disk icon) on the Editor Window toolbar and save the function with the name "rect." It is recommended that the file name be the same as the function name. An M-file with the name "rect.m" should appear in the Current Directory window. Close the Editor Window. The function "rect" can now be called by any other script that is operating within the "Fourier Optics" folder.

3.2 Creating Vectors

Suppose we want to program a simulation of a 1D spatial rectangle function with a half width of 0.055 m (full width of 0.11 m). This rectangle function will be created in a vector that corresponds to a physical length (side length L) of 2 m. If the number of samples in the vector is 200, then the sample interval Δx (or "dx" in the code) is 0.01 m. Open a new Editor Window by going to the main MATLAB desktop toolbar and clicking on New M-file. In the new window, enter the following lines:

```
1   % fft_example - Chapter 3 fft example
2
3   w=0.055;        %rectangle half-width (m)
4   L=2;            %vector side length (m)
5   M=200;          %number of samples
6   dx=L/M;         %sample interval (m)
```

Here, the parameters for the rectangle physical size, the vector side length, the number of samples, and sample interval are defined. Add the following code:

```
7   x=-L/2:dx:L/2-dx;   %coordinate vector
8   f=rect(x/(2*w));    %signal vector
```

The "*" indicates multiplication. In this code, the coordinate vector x is created with values ranging from −1 (= -L/2) to 0.99 (= L/2-dx) in steps of 0.01 (dx). The colons separate the range and step size. The subtraction of one step off the high limit results in a vector of 200 samples. The signal vector "f" is created with the help of the previously defined rect function and the use of the coordinate vector x. Since 2*w is a scalar, the interpretation is that each sample in x is divided by 2*w and the resulting vector is input to the rect function. To display what you have created in a plot of the f values against the x values, add the following lines:

```
9   figure(1)
10  plot(x,f);      %plot f vs x
```

The "figure(1)" command opens a window labeled "Figure 1" and "plot" makes a plot. Before executing this script, you need to save it to a disk. Click on Save on the Editor Window toolbar and save this M-file with the name fft_example. Don't use any spaces in the name, as that can be interpreted as a request to link to another function; to execute, click on Run in the Editor Window toolbar (a triangle icon or for older MATLAB versions an arrow beside a document icon). If there are no errors in your code, the plot of f should appear

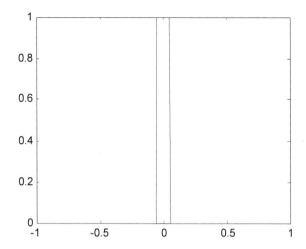

Figure 3.2 A rect function example plot.

in the Figure 1 window. If something is wrong with the code, you will see a message in the Command Window. The plot should look like Fig. 3.2.

There are many options for adjusting the plot display, which include changing the limits on the axes, adding labels, and changing graph lines and symbols. One approach is to use the "Edit Plot" (arrow icon) editor on the toolbar that appears at the top of each figure window. In this tutorial we will not go over the use of this editor and the reader is referred to the MATLAB "Help" (the "?" icon on the MATLAB Desktop window) to learn more.

The plot editor is a way to interactively work with the display. However, the changes made to the plot are not recorded in the code; so, the next time the script is executed, the plot will be displayed as it was before editing. Commands can be included in the script to adjust the plot. For our example, go back and edit the last `plot` command and add "`axis`" and "`xlabel`" command lines:

```
11  figure(2)
12  plot(x,f,'-o');   %plot f vs x
13  axis([-0.2 0.2 0 1.5]);
14  xlabel('x(m)');
```

The added option in the `plot` command (`'-o'`) places the marker o at each sample in the plot. The `axis` command sets limits on the values plotted on *x* and *y* axes and the `xlabel` command labels the *x*-axis. The `figure(2)` command will open a second window for this new plot. Without this command, the new plot would overwrite the first plot in the Figure 1 window. Note that the plot-modifying statements come after the plot call.

Click on Run to execute the code (this also automatically saves the latest version of the code) and the plot appearing in the Figure 2 window (Fig. 3.3) more clearly shows the rectangle function. Note the distance across the top of the rectangle corresponds to 0.1 m (11 samples) whereas the base of the rectangle corresponds to 0.12 m (13 samples). The correct interpretation for this sampled rectangle function is a width of 0.11 m, which lies between the two measures. In fact, because the rect function coding generates an odd number of samples, if a half width of 0.05 m were used rather than 0.055 m, the resulting signal vector would be identical to that shown in Fig. 3.3, which means there would be a small disparity in the intended width and the digital representation.

3.3 Shift for FFT

Disregarding the coordinate vector x for a moment, in terms of the 200 samples in f, a rectangle function with a width of 11 samples has been created that is centered in the middle of the vector f. This can be visualized by adding the following code:

```
15  figure(3)
16  plot(f,'-o');
17  axis([80 120 0 1.5]);
18  xlabel('index');
```

With the x argument removed from the plot command, the vector f is plotted as a function of the vector index values. The resulting plot (Fig. 3.4) shows the center sample of the rectangle function at index position 101. Centering the rectangle in the middle of the vector allows for easy viewing, but as described in Section 2.4, the FFT algorithm expects the zero-coordinate value to be in the first index location. To shift the values for the FFT operation, the "fftshift" command can be applied, which takes the samples of the first half of the vector and swaps them with the samples in the second half. Add the following the code to the program and run the script to get the plot in Fig. 3.5.

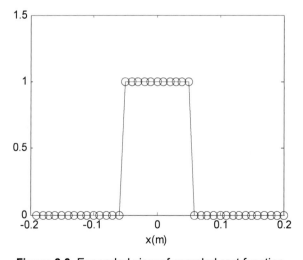

Figure 3.3 Expanded view of sampled rect function.

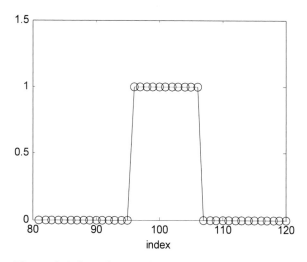

Figure 3.4 Sampled rect function versus index value.

```
19  f0=fftshift(f);    %shift f
20  figure(4)
21  plot(f0);
22  axis([0 200 0 1.5]);
23  xlabel('index');
```

A close look at Fig 3.5 should convince you that the center sample of the rectangle is at index location 1, and five samples are found to the "right" and the remaining five to the "left," where the left group is placed at the end of the vector.

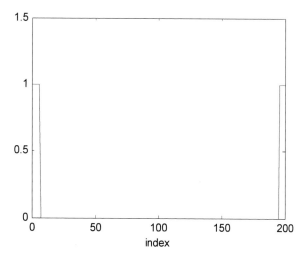

Figure 3.5 Shifted rect function versus index value.

A few comments about shifting for the FFT: first, it is more straightforward to use an even number of samples M for the vector and center the function of interest at an index of $M/2+1$. The `fftshift` function will then shift the $M/2+1$ sample to index 1. In our example $M = 200$ and the definition of the x vector causes the rectangle function to be centered at the $M/2+1$ position (101). More care needs to be taken when shifting vectors that contain an odd number of samples. For example, if the `fftshift` function is used with an odd number of samples, then the `ifftshift` function should be used to undo the shift. For an even number of samples, the `fftshift` function works both forward and backward. A second comment is that without the shift operation the FFT algorithm generates a transform for a function that is translated from the zero position, which means a linear phase term will be present in the result (shift theorem!).

3.4 Computing the FFT and Displaying Results

To compute the FFT of the vector "`f0`," add the following below the last piece of code:

```
24  F0=fft(f0)*dx;      %FFT and scale
25  figure(5)
26  plot(abs(F0));      %plot magnitude
27  title('magnitude');
28  xlabel('index');
29
30  figure(6)
31  plot(angle(F0));    %plot phase
32  title('phase');
33  xlabel('index');
```

Here, the 1D FFT algorithm in MATLAB is used. A capital letter is used for the frequency domain vector. Multiplying the result by the sample spacing `dx` is necessary to correctly approximate the analytic Fourier transform integral. Since each sample in `F0` contains two pieces of information (the real part and the imaginary part); or alternatively, the magnitude and the phase, two plots can be used to display this result. The "`abs`" command takes the absolute value (magnitude) of the samples in `F0` and the "`angle`" command extracts the phase of the samples, with output values ranging from $-\pi$ to π. Plot titles and x-axis labels have been included in this code. Run the script and the plots should look like those in Fig. 3.6.

MATLAB Programming of Functions, Vectors, Arrays, and Fourier Transforms 37

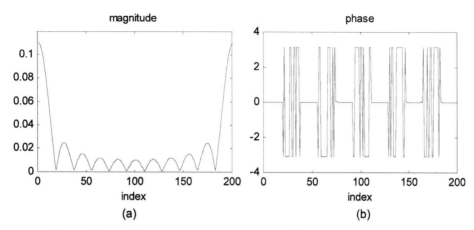

Figure 3.6 (a) Magnitude and (b) phase plots for FFT result versus index.

The sinc function nature of the magnitude result in Fig. 3.6 is obvious where the peak of the sinc is centered at index 1. However, note that the samples are not located exactly where the "zeros" of the magnitude would occur. Thus, the valleys shown in the curve do not necessarily appear to reach zero. The phase plot may look curious, but there are essentially three phase values in the plot: 0, $-\pi$, and π. However, $-\pi$ and π represent the same value in a modulo 2π format, so the sharp transitions in the plot between $-\pi$ and π are not particularly important, and the phase is effectively constant in these places. These sharp 2π transitions occur because of slight numerical differences between samples. With this interpretation, you should understand that the important phase transition is from 0 to π (or $-\pi$), which occurs about every 18 points. Comparing the phase plot with the magnitude plot, it is apparent that the π transition occurs at the magnitude zeros. Furthermore, a phase value of π is equivalent to placing a minus sign on the magnitude value. Combining all of this information, the magnitude and phase plots of Fig. 3.6 represent a real-valued sinc function where the values in the main lobe are positive, the values in the first lobe are negative, the second lobe values are positive, and so on. In this case the real-valued sinc function could have simply been displayed on one plot; but, in general, Fourier transform results are complex.

Once again for display reasons, it is helpful to center the FFT result in the vector. In addition, the spatial frequency coordinates need to be determined. Add the following to your script (cutting and pasting from the earlier code can help accomplish this quickly):

```
34  F=fftshift(F0);    %center F
35  fx=-1/(2*dx):1/L:1/(2*dx)-(1/L); %freq cords
36
37  figure(7)
38  plot(fx,abs(F));   %plot magnitude
39  title('magnitude');
```

```
40  xlabel('fx (cyc/m)');
41
42  figure(8)
43  plot(fx,angle(F));    %plot phase
44  title('phase');
45  xlabel('fx (cyc/m)');
```

The frequency coordinates in fx are created as defined in Section 2.4. Running the script generates the plots shown in Fig. 3.7, where the sampled Fourier magnitude and phase are centered, and the frequency axis is scaled appropriately.

3.5 Comparison with Analytic Results

When possible, it is good practice to test a new piece of code by applying simple inputs where the result can be compared with an analytic result. This helps diagnose problems and lets you build on previous code with confidence. For the example given in Section 3.4, an analytic result is easily found. For the expression

$$f(x) = \text{rect}\left(\frac{x}{2w}\right),$$

the Fourier transform is

$$F(f_X) = 2w\,\text{sinc}(2wf_X).$$

To compare the discrete result with the analytic result, add the following code to the "fft_example" script:

```
46  F_an=2*w*sinc(2*w*fx);    %analytic result
47
48  figure(9)
49  plot(fx,abs(F),fx,abs(F_an),':'); %plot magnitude
50  title('magnitude')
51  legend('discrete','analytic')
52  xlabel('fx (cyc/m)')
53
54  figure(10)
55  plot(fx,angle(F),fx,angle(F_an),':'); %plot phase
56  title('phase')
57  legend('discrete','analytic')
58  xlabel('fx (cyc/m)')
```

Here, the frequency coordinates fx are used as input to the analytic solution to create the vector F_an. Note that the MATLAB built-in sinc function is used since it conforms to our definition (see Table 1.2). The plot command

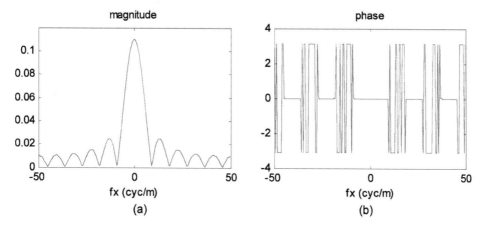

Figure 3.7 (a) Magnitude and (b) phase plots for centered FFT result.

arguments are set up to graph the discrete and analytic results on the same plot, and the legends are added to the display.

The resulting plots (Fig. 3.8) compare the FFT and analytic results. The magnitude results are nearly identical, but the FFT result has slightly higher values than the analytic curve at the edges. This is a result of the periodic extension property of the FFT (Section 2.5). The phase plots differ only in some transitions between π and $-\pi$ for the digital result, which are of no consequence.

3.6 Convolution Example

A convolution can be performed using the FFT and applying the Fourier convolution theorem. The example presented here involves the convolution of two Gaussian functions of different widths. In the Editor Window, select New M-file and enter the following code that defines and generates sample values for the two functions `fa` and `fb`:

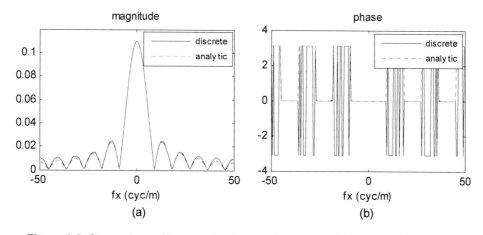

Figure 3.8 Comparison of (a) magnitude and (b) phase of FFT and analytic results.

```
1   % conv_example - Convolution: two Gaussian functions
2
3   wa=0.3;     %Gaussian 1 width [exp(-pi) radius](m)
4   wb=0.2;     %Gaussian 2 width [exp(-pi) radius](m)
5   L=2;        %side length (meters)
6   M=200;      %number of samples
7   dx=L/M;     %sample interval (m)
8
9   x=[-L/2:dx:L/2-dx];         %x coordinates
10  fa=exp(-pi*(x.^2)/wa^2);    %Gaussian a
11  fb=exp(-pi*(x.^2)/wb^2);    %Gaussian b
12
13  figure(1)
14  plot(x,fa,x,fb,'--'); title ('functions');
15  xlabel('x (m)');
```

Select the name "conv_example" for this M-file. Note the command for squaring each value in the x vector requires a period before the "^" symbol. Without the period, MATLAB will attempt a vector rather than a single element operation. Running this code produces the plot shown in Fig. 3.9(a). Now add the following to compute and plot the convolution result:

```
16  Fa=fft(fa);        %transform fa
17  Fb=fft(fb);        %transform fb
18  F0=Fa.*Fb;         %multiply pointwise
19  f0=ifft(F0)*dx;    %inverse transform and scale
20  f=fftshift(f0);    %center result
21
22  figure(2)
23  plot(x,f); title('convolution')
24  xlabel('x (m)')
```

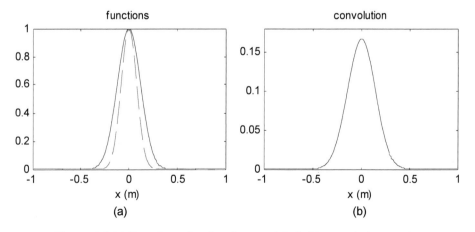

Figure 3.9 (a) Two Gaussian functions and their (b) convolution result.

The FFTs of the Gaussian vectors are first computed. The functions `fa` and `fb` need not be shifted prior to the FFTs since the convolution only depends on their relative positions. The transform results are next multiplied "pointwise" or element by element. This operation is indicated with a period placed before the product operator (.*). Without the period a vector/matrix multiplication would be attempted. Running the code produces the result shown in Fig. 3.9(b).

As introduced in Section 2.6, a critical consideration for a convolution computed in this way is the periodic extension property of the FFT. The criterion is that the sum of the function's supports should be less than the vector length. From Fig. 3.9(a), the estimated support of the significant values in the first Gaussian is $D_a \approx 0.9$ m and for the second $D_b \approx 0.6$, thus the sum of these two is less than the vector length $L = 2$ m. Exercise 3.2 gives you a chance to check if the curve in Fig. 3.9(b) is indeed a good representation of the analytic convolution.

In this book convolutions are coded directly by applying the convolution theorem, but MATLAB has the built-in function `conv` and for two dimensions, `conv2`. To strictly avoid artifacts due to the periodic convolution, these functions "zero-pad" the initial vectors, placing them in double-sized vectors or arrays of zeros, and then performing the convolution. For computational speed and efficiency we tend to work with fixed array sizes and pay heed to the support of the signals, as was done in the example above.

3.7 Two Dimensions

Physical optics problems typically involve at least two spatial dimensions. Start a new M-file, named fft2_example, and enter the following code to generate a two-dimensional (2D), sampled rectangle function:

```
1   % fft2_example - 2D FFT example
2
3   wx=0.1;              %rect x half-width (m)
4   wy=0.05;             %rect y half-width (m)
5   L=2;                 %side length x&y (m)
6   M=200;               %samples/side length
7   dx=L/M;              %sample interval (m)
8   x=-L/2:dx:L/2-dx;    %x coordinates
9   y=x;                 %y coordinates
10  [X,Y]=meshgrid(x,y); %X and Y grid coords
11  g=rect(X/(2*wx)).*rect(Y/(2*wy)); %signal
```

In this example, a 0.2×0.1 m rectangle is modeled in a 200×200 element array corresponding to a physical size of 2×2 m. The side length `L` is the physical length along one edge of the array where it is assumed that the x and y dimensions are the same. Two identical sample coordinate vectors `x` and `y` are defined for the two dimensions. The "`meshgrid`" command generates the coordinate arrays `X` and `Y` (capital letters) where the rows of `X` are copies of the vector x and the columns of Y are copies of y. X and Y are used to produce the sampled version of the 2D rect function in the array g. By using X and Y, the coded version of g appears much like an analytic expression.

An image is a common way to visualize optics simulation results. To display the contents of g as an image, add the following:

```
12  figure(1)
13  imagesc(x,y,g);        %image display
14  colormap('gray');      %linear gray display map
15  axis square;           %square figure
16  axis xy                %y positive up
17  xlabel('x (m)'); ylabel('y (m)');
```

The command "imagesc" scales the array data to the full display range and presents the image and "colormap" provides a pseudo-coloring of the image. The "gray" scale is used here, but check Help for other colormaps. The "axis square" command presents the figure border as a square rather than the default rectangular shape. This is helpful in this case where the side length is the same in the *x* and *y* directions. The first row of a conventional image file corresponds to the top of the picture. MATLAB image display functions assign *y*-axis values to increase from top to bottom. The axis xy command arranges the *y* axis to be displayed with increasing values from bottom to top. Running the script produces the image shown in Fig. 3.10(a).

It is also often helpful to view 1D "slices," or profiles, of 2D results. Add the following code to generate a 1D profile of the *x*-axis through the center of the array [Fig. 3.10(b)], where the syntax (M/2+1, :) selects the M/2+1 row:

```
18  figure(2)
19  plot(x,g(M/2+1,:));    %x slice profile
```

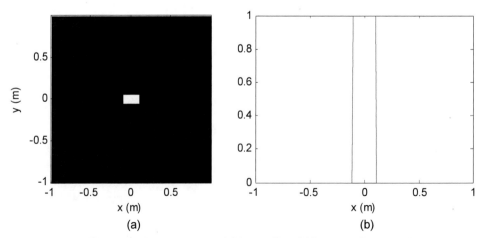

Figure 3.10 (a) Image and (b) *x* profile of 2D rectangle example.

```
20  xlabel('x (m)');
21  axis([-1,1,0,1.2]);
```

To perform a 2D FFT of the rectangle function, add the following:

```
22  g0=fftshift(g);        %shift
23  G0=fft2(g0)*dx^2;      %2D fft and dxdy scaling
24  G=fftshift(G0);        %center
25
26  fx=-1/(2*dx):1/L:1/(2*dx)-(1/L);%x freq coords
27  fy=fx;                 %y freq coords
```

The `fftshift` command operates on a 2D array in the manner described in Section 2.4 and the `fft2` command is needed for a 2D transform. Enter the following code to display the magnitude of the transform results as a surface plot along with a profile slice through the center (Fig. 3.11):

```
28  figure(3)
29  surf(fx,fy,abs(G))     %display transform magnitude
30  camlight left; lighting phong
31  colormap('gray')
32  shading interp
33  ylabel('fy (cyc/m)'); xlabel('fx (cyc/m)');
34
35  figure(4)
36  plot(fx,abs(G(M/2+1,:))); %fx slice profile
37  title('magnitude');
38  xlabel('fx (cyc/m)');
```

The "`surf`" plotting command is introduced to illustrate another display method for a 2D array. The `lighting`, `shading`, and `colormap` commands can be used to change the display. The magnitude could also be displayed in other ways, such as an image or a contour plot. The phase of the result is not shown here, but it could also be displayed in a variety of ways.

3.8 Miscellaneous Hints

A few other helpful MATLAB hints:

- Clear all: MATLAB retains the values of variables and arrays that are created when a script is executed. Typing a variable name in the Command Window and hitting enter displays the current value, which can be useful for analyzing your code. However, sometimes this gets confusing as code is being edited. The `clear all` command clears the variable memory.
- Close all: This command will close all the current figure windows.
- Display complex result: Sometimes a plot or image will not display and the error warning says the input is complex. After operations, like taking

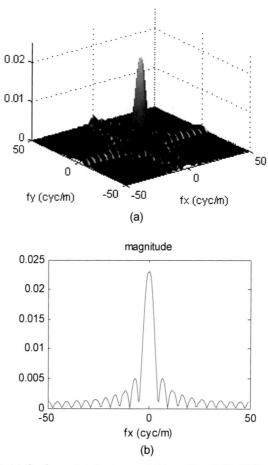

Figure 3.11 (a) Surface plot display and (b) profile of 2D FFT magnitude result.

the FFT, machine precision can leave some residual complex component in a vector or array that should be real. The `real` or `abs` functions can be applied to the array to allow the plot to display.

- Image contrast: The display function `imagesc` is used extensively throughout this book. For printing purposes the images are displayed in grayscale. However, it is easier to see low-value features with different colormaps—so try some other maps. To stretch the contrast of a grayscale image to more easily see dim features, a quick trick is to display the nth root of the image values. For example, `nthroot(g,3)` takes the third root of g. The higher the root, the more the contrast is stretched. Just be sure to remember that you are looking at a peak-scaled, contrast-stretched image.
- Vector operations: Pay special attention to vectorized operations indicated with the period—for example, an element-by-element multiplication indicated by `A.*B`. The most common programming error is to forget the period when a vectorized operation is needed. This mistake generally does

not create an error message, but the result will be unexpected. Make a habit of checking the vector operations when things are going wrong.
- 2D transform: Be sure and use `fft2` and `ifft2` for two dimensions—not `fft` and `ifft`. The 1D functions operating on the 2D array tend to give a repeated 1D result (it looks "striped")—which is not good!

3.9 Exercises

3.1 Triangle function M-file:

(a) Create a triangle function in an M-file. Try some lines like:
```
T=1-abs(x);
mask=abs(x)<=1;
out=T.*mask;
```

(b) In a script, create a sampled triangle function using the following specifications: triangle base half width = 0.1 m, vector length = 2 m, and number of samples $M = 200$.

(c) Plot the sampled function.

(d) Compute the FFT.

(e) Find the analytic Fourier transform of the function in (b).

(f) Plot the FFT and analytic Fourier transform results together (both magnitude and phase).

3.2 Code the example for the convolution of the Gaussian functions presented in Section 3.6. Find the analytic convolution of these functions and compare this result with the discrete result in a plot.

3.3 Circle function M-file:

(a) Create a circle function in an M-file.

(b) Generate a sampled circle function in a 2D array with the following parameters: circle radius = 0.015 m, array side length = 0.2 m, and number of samples (one dimension) $M = 200$.

(c) Display the sampled function as an image.

(d) Take the FFT of the array and display the magnitude of the transform in surface and profile plots.

Chapter 4
Scalar Diffraction and Propagation Solutions

Perhaps the most fundamental task associated with Fourier optics is describing the evolution of an optical field as it propagates from one location to another. The phenomenon of *diffraction* underlies the behavior of propagating waves. Extensive theory developed for diffraction provides the basis for modeling optical propagation on the computer. This chapter is essentially a summary of scalar diffraction theory with a listing of the expressions commonly used today to describe optical diffraction of monochromatic light. The presentation closely follows the diffraction development by Goodman.[1] More details can be found in that reference, as well as others.[2-4] This chapter sets the stage for the computer methods of simulating optical propagation described in Chapter 5.

4.1 Scalar Diffraction

Diffraction refers to the behavior of an optical wave when its lateral extent is confined; for example, by an aperture. It accounts for the fact that light rays do not follow strictly rectilinear paths when the wave is disturbed on its boundaries. In our everyday experience we rarely notice diffractive effects of light. The effects of reflection (from a mirror), or refraction (due to a lens) are much more obvious. In fact, the effects of diffraction become most apparent when the confinement size is on the order of the wavelength of the radiation. Nevertheless, diffraction plays a role in many optical applications and it is a critical consideration for applications involving high resolution, such as astronomical imaging, or long propagation distances such as laser radar, and in applications involving small structures such as photolithographic processes.

The propagation behavior of an optical wave is fundamentally governed by Maxwell's equations. In general, coupling exists between the wave's electric field $\vec{E}$ with components (E_x, E_y, E_z) and its magnetic field $\vec{H}$ with components (H_x, H_y, H_z). There is also coupling between the individual components of the electric field, as well as between the magnetic components. However, consider a wave that is propagating in a dielectric medium that is *linear* (field quantities from separate sources can be summed), *isotroptic* (independent of the wave polarization, i.e., the directions of $\vec{E}$ and $\vec{H}$), *homogeneous* (permittivity of the medium is independent of position), *nondispersive* (permittivity is independent of wavelength), and *nonmagnetic* (magnetic permeability is equal to the vacuum permeability). In this case, Maxwell's vector expressions become decoupled, and the behavior of each component of the electric or magnetic fields can be expressed independently from the other components. Scalar diffraction refers to the propagation behavior of light under this ideal situation.

The long list of assumptions for the medium suggests a rather limited application regime for scalar diffraction theory. However, scalar diffraction can clearly be used for describing free-space optical (FSO) propagation, which refers to transmission through space or the atmosphere and encompasses a huge number of interesting applications such as lidar, imaging, and laser communications. Furthermore, for many problems involving less benign propagation media, scalar solutions can provide a reasonable approximation of the principle effects of the propagation and establish a basis for comparison with full vector results. All of the developments and applications in this book assume scalar diffraction.

4.2 Monochromatic Fields and Irradiance

Some terminology and definitions related to optical fields are needed at this point. A *monochromatic* (single-frequency) scalar field propagating in an isotropic medium can be expressed as

$$u(P,t) = A(P)\cos[2\pi\nu t - \phi(P)], \qquad (4.1)$$

where $A(P)$ is the amplitude and $\phi(P)$ is the phase at a position P in space (x, y, z coordinates) and ν is the temporal frequency. This expression models a propagating transverse optical (electric) field of a single polarization.

Monochromatic light provides the basis for our analytic and computer simulation approaches to diffraction theory. A truly monochromatic light source is also *coherent*. Coherence refers to the correlation of the optical field phase at two different points in the field separated by time and/or space and enables the formation of interference in a time-averaged sense. Although some lasers can produce near-monochromatic radiation, true monochromatic light is unachievable. But, as will be discussed in Chapters 7 and 9, the extension of monochromatic results to polychromatic radiation, as well as partially coherent and incoherent radiation, can be straightforward in many useful cases (…fortunately!).

To give an example, a specific form of Eq. (4.1) corresponding to a plane wave propagating in the z direction would be

$$u(z,t) = A\cos[2\pi\nu t - kz], \qquad (4.2)$$

where the wavenumber k is defined as

$$k = \frac{2\pi}{\lambda}, \tag{4.3}$$

and where λ is the vacuum wavelength. Also, $\nu = c/\lambda$, where c is the speed of light in vacuum. This wave has no dependence on x and y and, therefore, is interpreted as extending infinitely in these directions.

If the field in Eq. (4.1) is propagating in a linear medium (assumed for scalar diffraction), the temporal frequency of the resulting field will remain unchanged; so, it is not necessary to explicitly carry the temporal term. Furthermore, substituting a complex *phasor* form for the cosine function provides a valid propagation result and aids in mathematical manipulation. These changes lead to a function that simply describes the spatial distribution of the field

$$U(P) = A(P)\exp[j\phi(P)]. \tag{4.4}$$

This complex phasor form of the optical field will be used extensively in our analytic and simulation developments. As an example, the phasor form of Eq. (4.2) is

$$U(z) = A\exp(jkz). \tag{4.5}$$

The descriptions in Eqs. (4.1) and (4.4) are related by

$$u(P,t) = \mathrm{Re}\{U(P)\exp(-j2\pi\nu t)\}, \tag{4.6}$$

where Re indicates the real part and the complex phasor $\exp(-j2\pi\nu t)$ is introduced for the temporal component of the field. To further refine Eq. (4.4), the explicit dependence on the z position can be removed, where z is assumed to be the fundamental propagation direction. Thus,

$$U_1(x,y) = A_1(x,y)\exp[j\phi_1(x,y)], \tag{4.7}$$

indicates the field in the x–y plane is located at some position "1" on the z axis.

Detectors do not currently exist that can follow the extremely high-frequency oscillations ($>10^{14}$ Hz) of the optical electric field. Instead, optical detectors respond to the time-averaged squared magnitude of the field. So, a quantity of considerable interest is the *irradiance*, which is defined here as

$$I_1(x,y) = U_1(x,y)U_1(x,y)^* = |U_1(x,y)|^2. \tag{4.8}$$

Irradiance is a radiometric term for the flux (watts) per unit area falling on the observation plane. It is a power density quantity that in other laser and Fourier optics references is often called "intensity." Expression (4.8) actually represents a shortcut for determining the time-averaged square magnitude of the field and is valid when the field is modeled by a complex phasor.

On a bookkeeping note, since $A_1(x,y)$ is the electric field amplitude, with typical units of volt/m, then to yield the corresponding irradiance value with units of watt/m^2 the right side of Eq. (4.8) needs to be multiplied by the constant $1/(2\eta)$ where η is the characteristic impedance of the medium ($\eta = 377\ \Omega$, for vacuum). Since we are most interested in the relative spatial form of the field, this constant is usually dropped in our discussions.

4.3 Optical Path Length and Field Phase Representation

The refractive index n of a medium is the ratio of the speed of light in vacuum to the speed in the medium. For example, a typical glass used for visible light might have an index of about 1.6. For light propagating a distance d in a medium of index n, the *optical path length* (OPL) is defined as

$$\text{OPL} = nd \ . \tag{4.9}$$

The OPL multiplied by the wavenumber k shows up in the phase of the complex exponential used to model the optical field. Think of k as the "converter" between the distance spanned by one wavelength and 2π rad of the phase. For example, in the plane-wave expression of Eq. (4.2), z is the OPL, where the propagation is assumed to be in vacuum; so, $n = 1$. The term kz gives the number of radians the sinusoid phase of the field has progressed over this distance. Sinusoids or complex exponentials are modulo 2π entities; so, only the relative phase between 0 and 2π has meaning. If the plane wave propagates a distance d through a piece of glass with index n, then the OPL is as indicated in Eq. (4.9), and the field phasor representation is

$$U(d) = A \exp(jknd). \tag{4.10}$$

In effect, the wavelength shortens to λ/n in the glass. There are other variations of this theme; for example, $\exp(jkr)$, where r is a radial distance in vacuum.

Phasor forms associated with the optical field can also be a function of transverse position x and y; for example,

$$\exp\left[j\frac{k}{2z}\left(x^2 + y^2\right)\right]. \tag{4.11}$$

This is known as a "chirp" term (see Appendix A) and indicates a field phase change as the square of the transverse position. This type of term appears in a

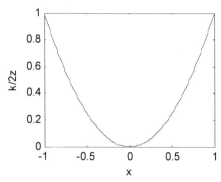

Figure 4.1 x-axis phase profile of the chirp function in Eq. (4.11).

variety of situations to model a contracting or expanding optical field. An example 1D profile of the phase of Eq. (4.11) is plotted in Fig. 4.1.

An important concept is *leading* and *lagging* phase. The temporal phasor $\exp(-j2\pi\nu t)$ defined in Eq. (4.6) indicates that the phase of the optical field becomes more *negative* as time progresses. Therefore, we say the phase in the center of the profile in Fig. 4.1 *leads* the rest of the function since it has the "most negative" value. The further away from the center, the more the phase *lags*. Interpreting the phase as a representation of an optical wavefront, the center of the wave crest in Fig. 4.1 leads the edges, and the wave can be imagined to be "propagating downward" in Fig. 4.1. Further physical interpretation of the optical phase is discussed in Section 5.3.

4.4 Analytic Diffraction Solutions

4.4.1 Rayleigh–Sommerfeld solution I

Consider the propagation of monochromatic light from a 2D plane (source plane) indicated by the coordinate variables ξ and η (Fig. 4.2). At the source plane, an area Σ defines the extent of a source or an illuminated aperture. The field distribution in the source plane is given by $U_1(\xi, \eta)$, and the field $U_2(x, y)$ in a distant observation plane can be predicted using the first *Rayleigh–Sommerfeld diffraction solution*

$$U_2(x,y) = \frac{z}{j\lambda} \iint_\Sigma U_1(\xi,\eta) \frac{\exp(jkr_{12})}{r_{12}^2} d\xi d\eta. \tag{4.12}$$

Here, λ is the optical wavelength; k is the wavenumber, which is equal to $2\pi/\lambda$ for free space; z is the distance between the centers of the source and observation coordinate systems; and r_{12} is the distance between a position on the source plane and a position in the observation plane. ξ and η are variables of integration, and the integral limits correspond to the area of the source Σ. With the source and observation positions defined on parallel planes, the distance r_{12} is

$$r_{12} = \sqrt{z^2 + (x-\xi)^2 + (y-\eta)^2} \,. \tag{4.13}$$

Expression (4.12) is a statement of the Huygens–Fresnel principle. This principle supposes the source acts as an infinite collection of fictitious point sources, each producing a spherical wave associated with the actual source field at any position (ξ, η). The contributions of these spherical waves are summed at the observation position (x, y), allowing for interference. The extension of Eqs. (4.12) and (4.13) to nonplanar geometries is straightforward; for example, involving a more complicated function for r, but the planar geometry is more commonly encountered, and this is our focus here.

Expression (4.12) is, in general, a superposition integral, but with the source and observation areas defined on parallel planes, it becomes a convolution integral, which can be written as

$$U_2(x,y) = \iint U_1(\xi,\eta) h(x-\xi, y-\eta) d\xi d\eta \,, \tag{4.14}$$

where the general form of the Rayleigh–Sommerfeld impulse response is

$$h(x,y) = \frac{z}{j\lambda} \frac{\exp(jkr)}{r^2}, \tag{4.15}$$

and

$$r = \sqrt{z^2 + x^2 + y^2} \,. \tag{4.16}$$

The Fourier convolution theorem is applied to write Eq. (4.14) as

$$U_2(x,y) = \Im^{-1}\{\Im\{U_1(x,y)\}\Im\{h(x,y)\}\}. \tag{4.17}$$

For this convolution interpretation the source and observation plane variables are

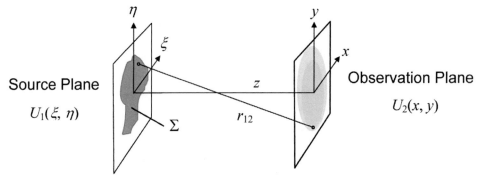

Figure 4.2 Propagation geometry for parallel source and observation planes.

simply re-labeled as x and y. An equivalent expression for Eq. (4.17) is

$$U_2(x,y) = \mathfrak{I}^{-1}\{\mathfrak{I}\{U_1(x,y)\}H(f_X, f_Y)\}, \tag{4.18}$$

where H is the Rayleigh–Sommerfeld transfer function given by

$$H(f_X, f_Y) = \exp\left(jkz\sqrt{1-(\lambda f_X)^2-(\lambda f_Y)^2}\right). \tag{4.19}$$

Strictly speaking, $\sqrt{f_X^2 + f_Y^2} < 1/\lambda$ must be satisfied for propagating field components. An angular spectrum analysis is often used to derive Eq. (4.19).

The Rayleigh–Sommerfeld expression is the most accurate diffraction solution considered in this book. Other than the assumption of scalar diffraction, this solution only requires that $r \gg \lambda$, the distance between the source and the observation position, be much greater than a wavelength.

4.4.2 Fresnel approximation

The square root in the distance terms of Eq. (4.13) or (4.16) can make analytic manipulations of the Rayleigh–Sommerfeld solution difficult and add execution time to a computational simulation. By introducing approximations for these terms, a more convenient scalar diffraction form is developed. Consider the binomial expansion

$$\sqrt{1+b} = 1 + \frac{1}{2}b - \frac{1}{8}b^2 + \ldots, \tag{4.20}$$

where b is a number less than 1, then expand Eq. (4.13) and keep the first two terms to yield

$$r_{12} \approx z\left[1 + \frac{1}{2}\left(\frac{x-\xi}{z}\right)^2 + \frac{1}{2}\left(\frac{y-\eta}{z}\right)^2\right]. \tag{4.21}$$

This approximation is applied to the distance term in the phase of the exponential in Eq. (4.12), which amounts to assuming a parabolic radiation wave rather than a spherical wave for the fictitious point sources. Furthermore, use the approximation $r_{12} \approx z$ in the denominator of Eq. (4.12) to arrive at the *Fresnel diffraction* expression:[1]

$$U_2(x,y) = \frac{e^{jkz}}{j\lambda z}\iint U_1(\xi,\eta)\exp\left\{j\frac{k}{2z}\left[(x-\xi)^2 + (y-\eta)^2\right]\right\}d\xi d\eta. \tag{4.22}$$

This expression is also a convolution of the form in Eq. (4.14), where the impulse response is

$$h(x, y) = \frac{e^{jkz}}{j\lambda z} \exp\left[\frac{jk}{2z}(x^2 + y^2)\right], \quad (4.23)$$

and the transfer function is

$$H(f_X, f_Y) = e^{jkz} \exp\left[-j\pi\lambda z(f_X^2 + f_Y^2)\right]. \quad (4.24)$$

The expressions in Eqs. (4.17) and (4.18) are again applicable in this case for computing diffraction results.

Another useful form of the Fresnel diffraction expression is obtained by moving the quadratic phase term that is a function of x and y outside the integrals:

$$U_2(x, y) = \frac{\exp(jkz)}{j\lambda z} \exp\left[j\frac{k}{2z}(x^2 + y^2)\right]$$
$$\times \iint \left\{U_1(\xi, \eta) \exp\left[j\frac{k}{2z}(\xi^2 + \eta^2)\right]\right\} \exp\left[-j\frac{2\pi}{\lambda z}(x\xi + y\eta)\right] d\xi d\eta.$$
$$(4.25)$$

Along with the amplitude and chirp multiplicative factors out front, this expression is a Fourier transform of the source field times a chirp function where the following frequency variable substitutions are used for the transform:

$$f_\xi \rightarrow \frac{x}{\lambda z}, \qquad f_\eta \rightarrow \frac{y}{\lambda z}. \quad (4.26)$$

The accuracy of the Fresnel expression when modeling scalar diffraction at close ranges suffers as a consequence of the approximations involved. By allowing a 1 rad maximum phase change [due to dropping the terms $b^2/8$ and above in the series of Eq. (4.20)], the following condition is derived:

$$z^3 \gg \left(\frac{\pi}{4\lambda}\left[(x-\xi)^2 + (y-\eta)^2\right]^2\right)_{\max}, \quad (4.27)$$

where the "max" notation indicates the maximum value that is of interest for a given source and observation plane geometry.

The criterion of Eq. (4.27) provides a well-defined condition, where the Fresnel approximation can be applied with little loss of accuracy. However, for fields in the source plane with little spatial variation, such as a simple aperture back-illuminated by a plane-wave, the Fresnel approximation can provide high accuracy even when Eq. (4.27) is violated. A looser criterion involves the *Fresnel number*, which is commonly used for determining when the Fresnel expression can be applied. The Fresnel number is given by

$$N_F = \frac{w^2}{\lambda z}, \qquad (4.28)$$

where w is the half width of a square aperture in the source plane, or the radius of a circular aperture, and z is the distance to the observation plane. If N_F is *less than ≈ 1* for a given scenario, then it is commonly accepted that the observation plane is in the *Fresnel region*, where the Fresnel approximations, typically, lead to useful results. However, for relatively "smooth" fields over the source aperture, the Fresnel expression can be applicable up to Fresnel numbers of even 20 or 30. In a geometrical optics context, the Fresnel expression describes diffraction under the paraxial assumption, where only rays that make a small angle ($< \sim 0.1$ rad) relative to the optical axis are considered.

4.4.3 Fraunhofer approximation

Fraunhofer diffraction, which refers to diffraction patterns in a regime that is commonly known as the "far field," is arrived at mathematically by approximating the chirp term multiplying the initial field within the integrals of Eq. (4.25) as unity. The assumption involved is

$$z \gg \left(\frac{k(\xi^2 + \eta^2)}{2}\right)_{max}, \qquad (4.29)$$

and results in the *Fraunhofer diffraction* expression:

$$U_2(x_2, y_2) = \frac{\exp(jkz)}{j\lambda z} \exp\left[j\frac{k}{2z}(x^2 + y^2)\right]$$
$$\times \iint U_1(\xi, \eta) \exp\left[-j\frac{2\pi}{\lambda z}(x\xi + y\eta)\right] d\xi d\eta. \qquad (4.30)$$

The condition of Eq. (4.29), typically, requires very long propagation distances relative to the source support size. However, a form of the Fraunhofer pattern also appears in the propagation analysis involving lenses. The Fraunhofer diffraction expression is a powerful tool and finds use in many applications such as laser beam propagation, image analysis, and spectroscopy.

Along with multiplicative factors out front, the Fraunhofer expression can be recognized simply as a Fourier transform of the source field with the variable substitutions

$$f_\xi \to \frac{x}{\lambda z}, \qquad f_\eta \to \frac{y}{\lambda z}. \qquad (4.31)$$

The Fraunhofer expression cannot be written as a convolution integral, so there is no impulse response or transfer function. But, since it is a scaled version of the Fourier transform of the initial field, it can be relatively easy to calculate, and as with the Fresnel expression, the Fraunhofer approximation is often used with success in situations where Eq. (4.29) is not satisfied. For simple source structures such as a plane-wave illuminated aperture, the Fraunhofer result can be useful even when Eq. (4.29) is violated by more than a factor of 10, particularly if the main quantity of interest is the irradiance pattern at the receiving plane. Using the Fresnel number N_F, the *commonly accepted requirement for the Fraunhofer region is $N_F \ll 1$.*

4.5 Fraunhofer Diffraction Example

It is extremely difficult (impossible?) to find closed-form diffraction solutions using the Rayleigh–Sommerfeld expression for most apertures. The Fresnel expression is more tractable, but solutions are still complicated even for simple cases such as a rectangular aperture illuminated by a plane wave.[1,2] So, Fresnel or Rayleigh–Sommerfeld calculations are left for the computer in the next chapter. Analytic Fraunhofer diffraction analysis is easier and, for our purposes, serves as a check on some of the computer results.

Consider a circular aperture illuminated by a unit amplitude plane wave. The complex field immediately beyond the aperture plane is

$$U_1(\xi,\eta) = \mathrm{circ}\left(\frac{\sqrt{\xi^2 + \eta^2}}{w}\right). \qquad (4.32)$$

To find the Fraunhofer diffraction field, the Fourier transform is taken as

$$\mathfrak{I}\{U_1(\xi,\eta)\} = w^2 \frac{J_1\left(2\pi w \sqrt{f_\xi^2 + f_\eta^2}\right)}{w\sqrt{f_\xi^2 + f_\eta^2}}. \qquad (4.33)$$

Then, with the substitutions in Eq. (4.31), and applying the leading amplitude and phase terms of Eq. (4.30), the field is found with

Scalar Diffraction and Propagation Solutions

$$U_2(x,y) = \frac{\exp(jkz)}{j\lambda z} \exp\left(j\frac{k}{2z}(x^2+y^2)\right)$$

$$\times w^2 \frac{J_1\left(2\pi \frac{w}{\lambda z}\sqrt{x^2+y^2}\right)}{\frac{w}{\lambda z}\sqrt{x^2+y^2}}. \qquad (4.34)$$

The irradiance, using Eq. (4.8), is

$$I_2(x,y) = \left(\frac{w^2}{\lambda z}\right)^2 \left[\frac{J_1\left(2\pi \frac{w}{\lambda z}\sqrt{x^2+y^2}\right)}{\frac{w}{\lambda z}\sqrt{x^2+y^2}}\right]^2. \qquad (4.35)$$

Some of the $w/\lambda z$ terms could be canceled, but the symmetry of this form is helpful for programming.

Let's exercise MATLAB to display this irradiance pattern. Suppose $w = 1$ mm and $\lambda = 0.633$ μm (He–Ne laser wavelength). The Fresnel number constraint requires $w^2/\lambda z < 0.1$ or $z > 10w^2/\lambda$, which leads to $z > 15.8$ m. We'll use $z = 50$ m.

Now to choose some mesh parameters. A good display size for the function is if the array side length is perhaps five times wider than the pattern's central lobe. The Bessel function J_1 has a first zero when the argument is equal to 1.22π. If $y = 0$, then the first zero in the pattern occurs when

$$2\pi \frac{w}{\lambda z} x = 1.22\pi. \qquad (4.36)$$

Solve for x to get half the center lobe width and double this result to get the full width of the center lobe

$$D_{lobe} = 1.22 \frac{\lambda z}{w}. \qquad (4.37)$$

We will choose $L = 5 \times 1.22 \lambda z/w \approx 0.2$ m.

Now for some code. It is helpful to first make a function that handles the Bessel function ratio. In a New M-file (named "jinc") enter the following:

```
1   function[out]=jinc(x);
2   %
3   % jinc function
4   %
5   % J1(2*pi*x)/x
```

```
6    % divide by zero fix
7    %
8    % locate non-zero elements of x
9    mask=(x~=0);
10   % initialize output with pi (value for x=0)
11   out=pi*ones(size(x));
12   % compute output values for all other x
13   out(mask)=besselj(1,2*pi*x(mask))./(x(mask));
14   end
```

This function evaluates $J_1(2\pi x)/x$. A masking approach is used to avoid the divide-by-zero condition when $x = 0$. The masking code may appear to be a roundabout way of doing things, but it allows the input x to be a vector or an array. In line 9, the array mask picks up the dimension of x and takes on a value of 1 for any element where x is nonzero (~= means ≠). In line 11, out is initialized with the dimension of x, ones fills the array with 1s, and π is the value of the function for $x = 0$. Then, logical indexing is applied—out(mask) and x(mask)—to evaluate the function for all elements where mask is 1. This leaves the value of π for $x = 0$. The MATLAB call besselj(1,...) is the Bessel function of the first kind, order 1.

As with sinc functions, there are several definitions in the literature for "jinc" functions, and this book may be the only one that uses this particular variation. So beware, not all jinc functions are the same. Now for the Fraunhofer pattern. Name this file "fraun_circ":

```
1    %fraun_circ - Fraunhofer irradiance plot
2
3    L=0.2;           %side length (m)
4    M=250;           %# samples
5    dx=L/M;          %sample interval
6    x=-L/2:dx:L/2-dx; y=x;   %coords
7    [X,Y]=meshgrid(x,y);
8
9    w=1e-3;          %x half-width
10   lambda=0.633e-6; %wavelength
11   z=50;            %prop distance
12   k=2*pi/lambda;   %wavenumber
13   lz=lambda*z;
14
15   %irradiance
16   I2=(w^2/lz)^2.*(jinc(w/lz*sqrt(X.^2+Y.^2))).^2;
17
18   figure(1)   %irradiance image
19   imagesc(x,y,nthroot(I2,3));
20   xlabel('x (m)'); ylabel('y (m)');
21   colormap('gray');
22   axis square;
23   axis xy;
```

Scalar Diffraction and Propagation Solutions 59

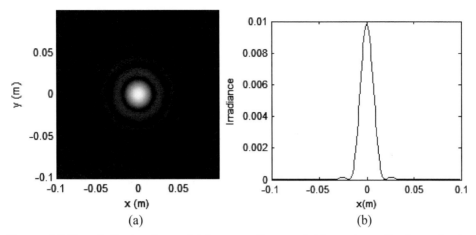

Figure 4.3 Fraunhofer irradiance (a) image pattern and (b) x-axis profile for a circular aperture. This is known as the Airy pattern.

```
24
25  figure(2)    %x-axis profile
26  plot(x,I2(M/2+1,:));
27  xlabel('x(m)'); ylabel('Irradiance');
```

Here are a few comments on this routine with associated line numbers:

(a) Line 6: The subscript 2 is left off the coordinate variables for simplicity.
(b) Line 9: Scientific notation can be done several ways: e-3 and 10^-3 mean the same thing. Don't use the ^ symbol in the exponential e notation!
(c) Line 16: jinc function is called.
(d) Line 19: 3rd root is used to bring out the "rings" in the image display.

Running the script produces the results in Fig. 4.3. The Fraunhofer pattern of a circular aperture is commonly known as the *Airy pattern*. The central core of this pattern, whose width is given in Eq. (4.37), is known as the *Airy disk*.

4.6 Exercises

4.1 Consider a plane wave of wavelength λ incident on two pieces of glass of different thicknesses and refractive indices as shown in Fig. 4.4. Find an expression for the optical path (length) difference (OPD) for the two parts of the beam between planes a and b.

4.2 For plane-wave illumination ($\lambda = 0.5$ μm) of an aperture of *diameter* 1 mm, determine the range of propagation distances that are adequate for the Rayleigh–Sommerfeld, Fresnel, and Fraunhofer diffraction regimes.

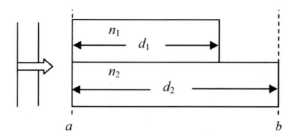

Figure 4.4 Plane wave propagation through two differing pieces of glass.

4.3 Verify that the Fresnel transfer function is the Fourier transform of the Fresnel impulse response.

4.4 Derive analytic expressions for the Fraunhofer field and irradiance patterns for the following apertures (shown in Fig. 4.5) illuminated with a unit amplitude plane wave. Plot the analytic Fraunhofer irradiance pattern images and profiles for the apertures on the computer. Choose suitable propagation distances z and side lengths L in the observation plane. Use $\lambda = 0.633$ μm and the following parameters:

(a) $w_X = 0.1$ mm, $w_Y = 0.05$ mm.

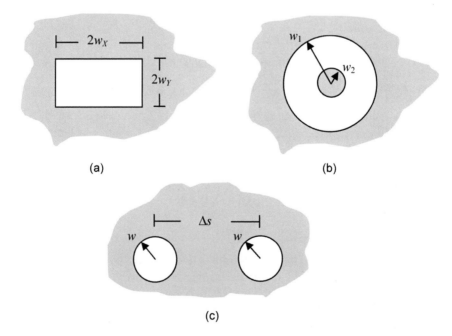

Figure 4.5 Apertures: (a) rectangle, (b) circle with obscuration, and (c) pair of circles.

(b) $w_1 = 1$ mm, $w_2 = 0.2$ mm.

(c) $w = 1$ mm, $\Delta s = 4$ mm.

4.5 Suppose the plane-wave field incident on the aperture of Fig. 4.5(c) is attenuated by different amounts in passing through each hole. The field exiting one hole has a magnitude of A_1 and the field exiting the other hole has a magnitude of A_2. Find an analytic expression for the Fraunhofer irradiance for this aperture. Plot the irradiance pattern and profile (along the *x*-axis) for $A_1 = 1$ and $A_2 = 0.4$. [Hint: take squared magnitude before combining complex exponentials into a cosine term.]

4.7 References

1. J. W. Goodman, *Introduction to Fourier Optics*, 3rd Ed., Roberts & Company, Greenwood Village, CO (2005).
2. J. D. Gaskill, *Linear Systems, Fourier Transforms, and Optics*, Wiley-Interscience, New York (1978).
3. O. K. Ersoy, *Diffraction, Fourier Optics, and Imaging*, Wiley-Interscience, New York (2006).
4. E. Hecht, *Optics*, 4th Ed., Addison-Wesley, Reading, MA (2002).

Chapter 5
Propagation Simulation

Now we look at several implementations of the diffraction expressions of Chapter 4 to simulate optical propagation. Although the material is presented as a teaching exercise, these propagation methods are used extensively in research and industry for modeling laser beam propagation. The concentration is on methods that use the fast Fourier transform (FFT) and only monochromatic light will be considered here. When designing a simulation there are a variety of issues related to discrete sampling that need to be considered. We will get to that, but let's first get our feet wet with some programming.

5.1 Fresnel Transfer Function (TF) Propagator

The Fresnel diffraction expression is often the approach of choice for simulations since it applies to a wide range of propagation scenarios and is relatively straightforward to compute. A common propagation routine is based on Eq. (4.18), which is repeated here for reference,

$$U_2(x,y) = \mathfrak{I}^{-1}\{\mathfrak{I}\{U_1(x,y)\}H(f_X,f_Y)\}, \qquad (5.1)$$

and uses the transfer function H given by

$$H(f_X,f_Y) = e^{jkz}\exp\left[-j\pi\lambda z\left(f_X^2 + f_Y^2\right)\right]. \qquad (5.2)$$

Start a New M-file and save it with name "propTF." Enter the following function:

```
1  function[u2]=propTF(u1,L,lambda,z);
2  % propagation - transfer function approach
3  % assumes same x and y side lengths and
4  % uniform sampling
5  % u1 - source plane field
6  % L - source and observation plane side length
7  % lambda - wavelength
8  % z - propagation distance
9  % u2 - observation plane field
```

```
10
11  [M,N]=size(u1);            %get input field array size
12  dx=L/M;                    %sample interval
13  k=2*pi/lambda;             %wavenumber
14
15  fx=-1/(2*dx):1/L:1/(2*dx)-1/L; %freq coords
16  [FX,FY]=meshgrid(fx,fx);
17
18  H=exp(-j*pi*lambda*z*(FX.^2+FY.^2));  %trans func
19  H=fftshift(H);             %shift trans func
20  U1=fft2(fftshift(u1));     %shift, fft src field
21  U2=H.*U1;                  %multiply
22  u2=ifftshift(ifft2(U2));   %inv fft, center obs field
23  end
```

This propagator function takes the source field u1 and produces the observation field u2 where the source and observation side lengths and sample coordinates are identical. Here are a few remarks on propTF with associated line numbers:

(a) Line 11: The size function finds the sample dimensions for the input field matrix u1 (only M is used). This helps reduce the number of parameters passed to the propTF function.
(b) Line 16: A line is saved by using fx twice in the meshgrid command since fy would be the same.
(c) Line 18: The transfer function H of Eq. (5.2) is programmed, although the exp(jkz) term is ignored. This term doesn't affect the transverse spatial structure of the observation plane result.
(d) Line 19: H is created in the array center but is shifted (fftshift) before the FFT.
(e) Line 20: Similarly, the source field u1 is assumed to be in the array center; so, fftshift is applied before the 2D FFT is computed.
(f) Line 21: U1 is multiplied "pointwise" by the transfer function H and the inverse FFT is computed to complete the convolution.
(g) Line 22: Finally, ifftshift centers u2 for display.

Note that lower case u is used for the spatial field and upper case U is used for Fourier domain quantities, which is not consistent with the use of upper case for the analytic spatial fields; for example, in Eq. (5.1). But what can you do? Both are established notations, so we live with a little notational mixing.

Before we exercise this function, let's look at another method to simulate propagation.

5.2 Fresnel Impulse Response (IR) Propagator

A propagation approach can be devised based on Eq. (4.17), which is repeated here:

$$U_2(x,y) = \mathfrak{I}^{-1}\{\mathfrak{I}\{U_1(x,y)\}\mathfrak{I}\{h(x,y)\}\}. \tag{5.3}$$

The impulse response h is given by

$$h(x,y) = \frac{e^{jkz}}{j\lambda z}\exp\left[\frac{jk}{2z}(x^2+y^2)\right]. \tag{5.4}$$

Even though Eqs. (5.1) and (5.3) represent identical analytical operations, with discrete sampling and transforms, the transfer function and impulse response approaches can yield different results.

For making the impulse response propagator, some typing can be saved by starting with a copy of propTF. For example, with propTF in the Editor, click on "File" on the MATLAB desktop window, select "Save As," edit the file name to be propIR, and click "Save." The new copy is now ready to be edited (carefully) to match the following:

```
1   function[u2]=propIR(u1,L,lambda,z);
2   % propagation - impulse response approach
3   % assumes same x and y side lengths and
4   % uniform sampling
5   % u1 - source plane field
6   % L - source and observation plane side length
7   % lambda - wavelength
8   % z - propagation distance
9   % u2 - observation plane field
10
11  [M,N]=size(u1);           %get input field array size
12  dx=L/M;                   %sample interval
13  k=2*pi/lambda;            %wavenumber
14
15  x=-L/2:dx:L/2-dx;         %spatial coords
16  [X,Y]=meshgrid(x,x);
17
18  h=1/(j*lambda*z)*exp(j*k/(2*z)*(X.^2+Y.^2));  %impulse
19  H=fft2(fftshift(h))*dx^2; %create trans func
20  U1=fft2(fftshift(u1));    %shift, fft src field
21  U2=H.*U1;                 %multiply
22  u2=ifftshift(ifft2(U2));  %inv fft, center obs field
23  end
```

This code is quite similar to the propTF function. Again, the source and observation planes in this approach have the same side length. Some specific remarks for propIR are as follows:

(a) Line 18: h is implemented, and the FFT is applied to get H.

(b) Line 19: Note the multiplier dx^2 for H. The FFT of u1 and FFT^{-1} of U2 take care of each other's scaling, but the FFT of h needs its own scaling.

Due to computational artifacts, the IR approach turns out to be more limited in terms of the situations where it should be used than the TF approach, however, it provides a way to simulate propagation over longer distances and is useful for the discussion of simulation limitations and artifacts.

5.3 Square Beam Example

Now it is time to try out the TF and IR propagators. Consider a source plane with dimensions 0.5 m × 0.5 m ($L_1 = 0.5$ m). Start a New M-file and use the name "sqr_beam." Enter the following:

```
1   % sqr_beam propagation example
2   %
3   L1=0.5;                %side length
4   M=250;                 %number of samples
5   dx1=L1/M;              %src sample interval
6   x1=-L1/2:dx1:L1/2-dx1; %src coords
7   y1=x1;
```

Variables with "1" are source plane quantities. The source and observation plane side lengths are the same for the TF and IR propagators, i.e., $L_1 = L_2 = L$, but this is not true for other propagators, so the variable L1 is retained here. The code sets up 250 samples across the linear dimension of the source plane, and the sample interval dx1 works out to be 2×10^{-3} m (2 mm).

Assume a square aperture with a half width of 0.051 m (51 mm) illuminated by a unit-amplitude plane wave from the backside, where $\lambda = 0.5$ μm. The simulation, therefore, places 51 samples across the aperture, which provides a good representation of the square opening (see Section 2.2). Add the following code:

```
8    lambda=0.5*10^-6;     %wavelength
9    k=2*pi/lambda;        %wavenumber
10   w=0.051;              %source half width (m)
11   z=2000;               %propagation dist (m)
12
13   [X1,Y1]=meshgrid(x1,y1);
14   u1=rect(X1/(2*w)).*rect(Y1/(2*w)); %src field
15   I1=abs(u1.^2);        %src irradiance
16   %
17   figure(1)
18   imagesc(x1,y1,I1);
19   axis square; axis xy;
20   colormap('gray'); xlabel('x (m)'); ylabel('y (m)');
21   title('z= 0 m');
```

Propagation Simulation

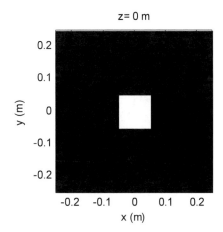

Figure 5.1 Source plane irradiance for sqr_beam propagation simulation.

The propagation distance z is 2000 m, so the Fresnel number for this case is $N_F = w^2/\lambda z = 2.6$, which is reasonable for applying the Fresnel expression. The source field is defined in the array u1. The irradiance is found by squaring the absolute value of the field. Executing this script generates Fig. 5.1, which verifies the source plane arrangement.

The next part of the script is where the propagation takes place. Add the following:

```
22  u2=propTF(u1,L1,lambda,z);   %propagation
23
24  x2=x1;                %obs coords
25  y2=y1;
26  I2=abs(u2).^2;        %obs irrad
27
28  figure(2)             %display obs irrad
29  imagesc(x2,y2,I2);
30  axis square; axis xy;
31  colormap('gray'); xlabel('x (m)'); ylabel('y (m)');
32  title(['z= ',num2str(z),' m']);
33  %
34  figure(3)             %irradiance profile
35  plot(x2,I2(M/2+1,:));
36  xlabel('x (m)'); ylabel('Irradiance');
37  title(['z= ',num2str(z),' m']);
38  %
39  figure(4)             %plot obs field mag
40  plot(x2,abs(u2(M/2+1,:)));
41  xlabel('x (m)'); ylabel('Magnitude');
42  title(['z= ',num2str(z),' m']);
43  %
44  figure(5)             %plot obs field phase
45  plot(x2,unwrap(angle(u2(M/2+1,:))));
```

```
46  xlabel('x (m)'); ylabel('Phase (rad)');
47  title(['z= ',num2str(z),' m']);
```

Observation plane parameters are indicated with a "2." In lines 24 and 25, the observation coordinates are set equal to the source coordinates since this is consistent with the operation of propTF. Lines 28–47 contain code to display an image of the observation plane irradiance as well as irradiance and field magnitude and phase profiles. The `num2str` function (line 47) is introduced to display the propagation distance in the plot title. The use of the `unwrap` function for the phase profile display (line 45) is discussed in the next few paragraphs.

Execute `sqr_beam`. The irradiance results are shown in Fig. 5.2, where the constructive and destructive interference of the coherent light, brought on by diffraction, results in peaks that have greater irradiance than the initial source value (1). The peaks are offset by valleys with less irradiance. Also apparent in the profile plot is a spread of the light into "wings" that exceed the initial width of the source rectangle.

The field magnitude and phase are shown in Fig. 5.3. The phase is in units of radians. The function `unwrap`, introduced for displaying the phase, adds multiples of $\pm 2\pi$ when jumps between consecutive samples are greater than π. This serves to remove the discontinuous skips that appear in the modulo-2π formatted phase function and helps the underlying form of the phase to be seen. Remove the `unwrap` command to see what the phase looks like with modulo-2π jumps.

In other words, Fig. 5.3(b) is representative of a slice through the surface of the constant phase for the optical wave at the observation plane. The important physical interpretation is that it represents the *shape of the optical wavefront* at the observation plane. Recall that the field temporal phasor, defined as $\exp(-j2\pi\nu t)$ in Eq. (4.6), is such that as time increases the phase becomes more negative. Therefore, the wavefront profile in Fig. 5.3(b), which is bowl shaped with a flattened center, can be thought of as progressing "downward" as time carries on. Furthermore, imagine rays projecting normal from the wavefront surface to get an idea of where the energy along the wavefront is headed. The magnitude plot in Fig. 5.3(a) shows that most of the field magnitude is concentrated near the flattened center of the wavefront and the phase indicates much of that energy is still headed normal to the observation plane.

Now try the impulse response IR propagator. Change line 22 to

```
u2=propIR(u1,L1,lambda,z);   %propagation
```

and run `sqr_beam`. The results in this case should be identical to those in Figs. 5.2 and 5.3.

Propagation Simulation 69

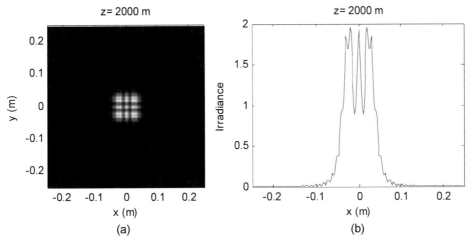

Figure 5.2 Observation plane irradiance (a) pattern and (b) profile.

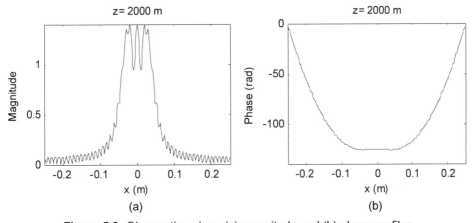

Figure 5.3 Observation plane (a) magnitude and (b) phase profiles.

5.4 Fresnel Propagation Sampling

5.4.1 Square beam example results and artifacts

Now for the bad news. Discrete sampling of the source field, sampling of the transfer function or impulse response, and the periodic nature of the FFT can lead to a variety of artifacts in the propagation result. Much of the trouble comes because the chirp functions on the right side of Eqs. (5.2) and (5.4) are not bandlimited and cannot be adequately sampled. This issue is introduced here with some example results.

In Fig. 5.4 both `propTF` and `propIR` results are shown for the `sqr_beam` routine at propagation distances ranging from 1000 to 20,000 m. At the distance of 1000 m the TF result appears reasonable with some constructive/destructive interference features and slight spreading beyond the initial aperture width. On the other hand, the IR result exhibits periodic copies of the pattern. At 2000 m

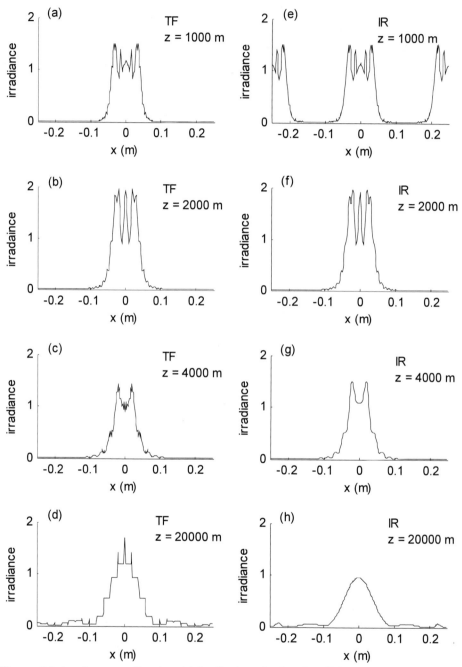

Figure 5.4 Irradiance profiles (*x* axis) for the transfer function (TF) and impulse response (IR) propagation approaches for propagation distances ranging from 1000 to 20,000 m. "Spiky" and "stair-step" artifacts appear in (c) and (d); periodic copies appear in (e); and spurious sidelobes appear in (h).

the TF and IR results are essentially identical. At 4000 m the TF result displays some "ticks" in the profile that are absent from the IR result. For 20,000 m the TF profile has a "stair-step" appearance that is clearly an artifact. The IR result is smooth. At longer distances the irradiance pattern is predicted by Fraunhofer theory to take on a sinc^2 form. This generally appears to be the case for Fig. 5.4(h). Corresponding irradiance patterns are shown in Fig. 5.5.

5.4.2 Sampling regimes and criteria

It turns out that Figs. 5.4(c)–(e) and 5.5(c)–(e) display significant artifacts related to discrete sampling. In this section we examine criteria used to predict when there will be problems. For more details on these and other criteria, see Appendix A and References 1 and 2 in this chapter (and A).

An obvious first criterion is that the support of the source field should "fit" within the numerical array. If D_1 is the effective support of the source field (maximum linear width) and L is the array side length, then we usually require

$$D_1 < L. \tag{5.5}$$

It is good practice to provide a "guard area" around the source function, for example, $L \approx 2D_1$ or $3D_1$. This helps reduce artifacts at the edges of the array after propagation due to the periodic extension properties of the FFT.

Further criteria are derived by considering the effects of sampling the chirp functions in the Fresnel transfer function H and impulse response h expressions (Appendix A). The H chirp, given as $\exp[-j\pi\lambda z(f_X^2 + f_Y^2)]$, is adequately sampled (oversampled) when

$$\Delta x \geq \frac{\lambda z}{L}. \tag{5.6}$$

This relationship is derived by considering aliasing of the chirp function in the frequency domain and then converting to the sampled space domain (Appendix A). The h chirp, given as $\exp[jk(2z)^{-1}(x^2 + y^2)]$, is oversampled when

$$\Delta x \leq \frac{\lambda z}{L}. \tag{5.7}$$

Expressions (5.6) and (5.7) reveal that the oversampling criteria for the transform pair H and h are opposite. Oversampling is a good thing, in general. If Eq. (5.6) or (5.7) is violated, an aliased representation of the phase of these respective chirp functions is created that leads to simulation artifacts. Both conditions are only satisfied when

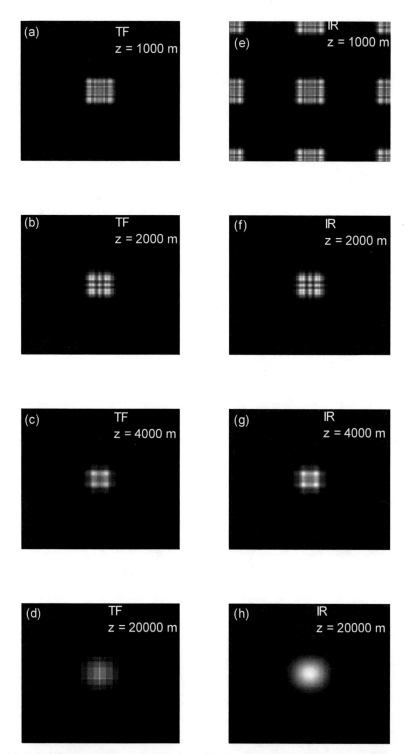

Figure 5.5 Irradiance patterns for the transfer function (TF) and impulse response (IR) propagation approaches corresponding to the profiles in Fig. 5.4.

$$\Delta x = \frac{\lambda z}{L}, \quad (5.8)$$

which we call the *critical sampling* condition.

Table 5.1 lists the three regimes defined by over-, under-, or critical sampling of the phase term in the H or the h functions. For each regime, a criterion is described that involves the source field bandwidth B_1. In practice, the source bandwidth criteria of Table 5.1 will never be satisfied because, as described in Section 2.2, a source with finite support will have infinite bandwidth. So, an effective bandwidth B_1 can be used when considering the criteria. Further comments on the sampling regimes are as follows:

(a) $\Delta x > \lambda z/L$: The "short distance" regime. Here, the support size available in the observation plane is limited. For the TF approach the size of the field in the observation plane that can be represented accurately is roughly $D_1 + \lambda z/\Delta x$ (see Appendix A). This limitation is usually not a problem, as the observation plane field is often negligible beyond the full width of $D_1 + \lambda z/\Delta x$. Thus, the TF

Table 5.1 Propagator sampling regimes and sampling criteria.

Regime	Chirp phase sampling	Source bandwidth criterion	Approach	Comments
$\Delta x > \dfrac{\lambda z}{L}$	TF: Over IR: Under	$B_1 \leq \dfrac{1}{2\Delta x}$	TF: Preferred IR: Periodic copies	Relatively "short" z or small λ Observation plane limitation TF: Observation plane field limited to full width $D_1 + \lambda z / \Delta x$
$\Delta x = \dfrac{\lambda z}{L}$	TF: Critical IR: Critical	$B_1 \leq \dfrac{1}{2\Delta x}$ or $B_1 \leq \dfrac{L}{2\lambda z}$	TF and IR identical	Full use of array space and bandwidth allocation
$\Delta x < \dfrac{\lambda z}{L}$	TF: Under IR: Over	$B_1 \leq \dfrac{L}{2\lambda z}$	TF: If bandwidth criterion essentially satisfied IR: Perhaps better if bandwidth criterion violated	Relatively "long" z or large λ Source bandwidth limitation

Note: Δx is the sample interval; B_1 is the source bandwidth; λ is the wavelength; D_1 is the support of the field in the source plane; and z is the propagation distance.

approach generally provides good results in this regime. The undersampled IR phase function has an aliased, periodic phase representation, and using this approach produces periodic copies of the field. The source bandwidth B_1 is only limited in the usual way by the sampling theorem in the source plane.

(b) $\Delta x = \lambda z/L$: This is the critical sampling situation where, remarkably, the sampled H and h functions as an FFT pair, turn out to have values that exactly match the analytic functions H and h (Appendix A). Under this condition the full bandwidth of the sampled array ($1/2\Delta x$) is available for modeling the source, and the full area of the array in the observation plane can be used.

(c) $\Delta x < \lambda z/L$: The "long distance" regime. Here, the bandwidth available for the source field becomes limited. The H chirp has an aliased phase form, where any significant source bandwidth that extends beyond $L/(2\lambda z)$ leads to artifacts using the TF approach. Applying the IR approach actually corresponds to windowing, or filtering the source frequency content beyond $\sim L/(2\lambda z)$. This leads to "smoother," but not always accurate, results.

5.4.3 Criteria applied to square beam example

What do the criteria of the previous section predict for the sqr_beam example of Section 5.3? The source field for that example is given by

$$U_1(x, y) = \text{rect}\left(\frac{x}{2w}\right)\text{rect}\left(\frac{y}{2w}\right), \tag{5.9}$$

where $w = 0.051$ m. The source support is $D_1 = 2w = 0.102$ m, which easily fits within the side length $L = 0.5$ m. This was illustrated in Fig. 5.1. To consider the criteria in Table 5.1, an estimate is needed for the source effective bandwidth. Referring to Section 2.2, a reasonable estimate for the effective bandwidth is

$$B_1 \approx \frac{5}{w} = 98 \text{ cycles/m}. \tag{5.10}$$

Table 5.2 presents pertinent sampling information for the four propagation distances in the sqr_beam example. Referring to Table 5.2 as well as Fig. 5.4, the following observations can be made:

(a) $z = 1000$ m: H is oversampled by a factor of 2 relative to critical sampling. However, the observation plane size limitation has a negligable effect on the TF result [Fig. 5.4(a)], as the most significant part of the field fits within $D_1 + \lambda z/\Delta x = 0.352$ m. The IR approach introduces periodic copies of the field separated by $\lambda z/\Delta x$ [Fig. 5.4(e)].

(b) $z = 2000$ m: Critical sampling. The TF and IR results are identical [Figs. 5.4(b) and (f)].

(c) $z = 4000$ m: H is undersampled by a factor of 2. The source bandwidth limit for propagation is 125 cyc/m, which is slightly larger than the source effective bandwidth B_1. Thus, most of the significant source spectrum obeys the criterion. But, small ticks still creep into the TF approach result [Fig. 5.4(c)]. On the other hand, artifacts are not apparent in the IR result [Fig. 5.4(h)].

(d) $z = 20,000$ m: H is undersampled by a factor of 10. The available bandwidth of 25 cycles/m is only a fourth of B_1. Thus, the TF approach causes significant stair-step artifacts [Fig. 5.4(d)]. The IR approach actually suppresses source frequency components that lie beyond the available bandwidth. This gives a smoother result, but with the small, spurious sidelobes near the array edge [Fig. 5.4(h)].

5.4.4 Propagator accuracy

How closely do the sqr_beam propagator examples follow the analytic Fresnel propagation result? Three particular cases are studied where the irradiance results appear to be "reasonable": TF at $z = 1000$ m, TF at $z = 2000$ m, and IR at $z = 20,000$ m. For comparison, an analytic result for a square aperture is available and is given by[3]

$$U_2(x,y) = \frac{e^{jkz}}{2j}\{[C(\alpha_2)-C(\alpha_1)] + j[S(\alpha_2)-S(\alpha_1)]\} \\ \times \{[C(\beta_2)-C(\beta_1)] + j[S(\beta_2)-S(\beta_1)]\}, \quad (5.11)$$

where C and S are known as the Fresnel integrals,

$$C(\chi) = \int_0^\chi \cos\left(\frac{\pi}{2}t^2\right)dt, \quad S(\chi) = \int_0^\chi \sin\left(\frac{\pi}{2}t^2\right)dt, \quad (5.12)$$

Table 5.2 Sampling regimes for the sqr_beam example.

Z (m)	$\frac{\lambda z}{\Delta x\, L}$	TF (H) sampling	IR (h) sampling	Available Observation plane size (m)	Source bandwidth limit (cycles/m)
1000	0.5	over	under	$D_1 + \lambda z/\Delta x = 0.352$	$1/2\Delta x = 250$
2000	1	critical	critical	$L = 0.5$	$1/2\Delta x = 250$
4000	2	under	over	$L = 0.5$	$L/2\lambda z = 125$
20,000	10	under	over	$L = 0.5$	$L/2\lambda z = 25$

Note: $\Delta x = 2$ mm; $L = 0.5$ m; $\lambda = 0.5$ μm; $N = 250$; and $D_1 = 0.102$ m.

and

$$\alpha_1 = -\sqrt{2/(\lambda z)}(w+x), \quad \alpha_2 = \sqrt{2/(\lambda z)}(w-x),$$

$$\beta_1 = -\sqrt{2/(\lambda z)}(w+y), \quad \beta_2 = \sqrt{2/(\lambda z)}(w-y). \quad (5.13)$$

Expression (5.11) was evaluated for the parameter values in the three propagator examples. Only the x-axis profile was considered ($y = 0$), and 2500 points were evaluated as opposed to 250 points in the simulation. The higher sample rate shows whether any spatial details are lost in the propagator results. A Fresnel integral routine was used to compute C and S. The MATLAB Symbolic Math Toolbox (available in the student edition package) has these functions, which are called "`mfun('FresnelC',x)`" and "`mfun('FresnelS',x)`." There are also other shared versions of Fresnel integral routines that are downloadable on the internet.

In Fig. 5.6 the field magnitude and unwrapped phase at the observation plane are compared with the analytic profiles. The TF and IR propagator profiles are displayed with solid lines, and the analytic results are displayed with dashed lines. There is generally good consistence between the curves. Note the magnitude results are plotted on a log scale to emphasize the small differences in the wings of the profiles. Of course, these results are specific to the `sqr_beam` example, but they give an indication of the typical performance of the FFT propagators. The following are comments on the propagator/analytic comparisons:

(a) TF, $z = 1000$ m: The significant features of the magnitude profiles [Fig. 5.6(a)] match, but smoothing is seen in the wings of the propagator result (the highly oscillating wings are the analytic result). The phase profiles [Fig. 5.6(b)] are nearly identical except at the edges of the array where the propagator phase slightly lags. $D_1 + \lambda z/\Delta x = 0.352$ m, which is roughly the apparent width of the magnitude curve before the propagator wings drop abruptly below the analytic curve. Overall, the propagator result appears quite accurate. The primary deviation is in the wings and is of little consequence.

(b) TF, $z = 2000$ m: The propagator magnitude curve [Fig. 5.6(c)] is nearly identical to the analytic curve, although the propagator curve is slightly elevated near the edge of the array as forced by periodic extension [difficult to see in Fig. 5.6(c)]. The phase profiles [Fig. 5.6(d)] are essentially identical. In this case critical sampling leads to an extremely close match with the analytic result.

(c) IR, $z = 20{,}000$ m: For this long-distance case, the IR propagator magnitude [Fig 5.6(e)] follows the analytic result but exhibits the spurious lobes near the array edges. The propagator phase in Fig. 5.6(f) also has errors in the wings. A better result could be found by readjusting

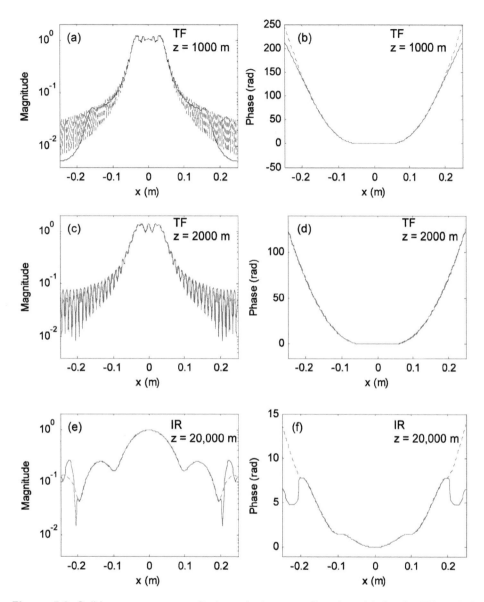

Figure 5.6 Solid curves are magnitude and phase profiles (*x* axis) for the TF and IR propagation approaches for propagation distances of 1000, 2000, and 20,000 m; dashed curves are corresponding analytic results.

the sampling parameters, but this irradiance pattern might be acceptable in a pinch.

5.4.5 Sampling decisions

Suggested steps for designing a propagation simulation are summarized in Table 5.3. As mentioned previously, a field with finite support in the source plane cannot be bandlimited, so the bandwidth criteria in Table 5.1 are never satisfied. However, the criteria can still be used to help find reasonable simulation parameters, although, it often becomes something of an art form to juggle sampling and field parameters to get a satisfactory propagation result.

Critical sampling helps minimize artifacts by allowing full use of the array side length and sampling bandwidth. It seems prudent to try and use critical sampling, but maintaining this condition can be inconvenient. Consider that the critical sampling expression $\Delta x = \lambda z / L$ can be rearranged to give

$$M = \frac{\lambda z}{\Delta x^2} = \frac{L^2}{\lambda z}, \quad (5.14)$$

which defines the critical sampling criterion in terms of the number of samples that span the array side length. For a given situation, the critical condition may dictate either too many samples for a practical FFT calculation or too few to adequately sample the source or observation planes. Other requirements can be at odds with the critical criterion. For example, phase screens used to simulate propagation through atmospheric turbulence have their own set of sample interval and array size conditions.[2,4] The sampling criterion can always be maintained with the help of interpolation or decimation; however, extra computational steps add complexity and run time and still may not solve the practical issues of working with too many, or too few, samples.

In practice, Step 3 in Table 5.3 is not always straightforward because it may not be simple to determine the effective bandwidth of the source. Sometimes the most expedient thing to do is make an "educated" guess at the sampling parameters and run a few trial simulations to see what happens. If there are signs of artifacts such as the stair-step or sidelobe features illustrated in Figs. 5.4 and 5.5, then the simulation parameters need a closer look.

5.4.6 Split-step simulation, windowing, and expanding grids

Some of you might say, "Hey, wait! Let's just break that long distance propagation problem into a sequence of shorter, better-behaved, TF propagations where each satisfies $\Delta x \geq \lambda z / L$." But, alas, the result is the same whether a single TF propagation or a sequence of shorter TF propagations is used. This is because a succession of TF propagations is the same as applying the product of the transfer functions to the initial field. For example, if the distance z_N is broken into a series of shorter distances, it is easy to show

$$H(f_X, f_Y; z_N) = H(f_X, f_Y; z_1) H(f_X, f_Y; z_2 - z_1) ... H(f_X, f_Y; z_N - z_{N-1}), \quad (5.15)$$

Table 5.3 Propagation simulation design steps.

(1) Consider source support: $D_1 < L$ by a factor of 2 or 3?

(2) Determine sampling regime (Table 5.2).

$\Delta x > \lambda z/L$: TF approach will often work well with some loss of observation plane support.

$\Delta x = \lambda z/L$: Critical sampling, TF approach, best use of bandwidth, and spatial support.

$\Delta x < \lambda z/L$: IR approach with loss of available source bandwidth, artifacts.

(3) Consider source bandwidth criteria in Table 5.2.

(4) Reconsider source sampling depending on bandwidth criteria and resulting artifacts.

---

where the z arguments are understood to replace z in Eq. (5.2). So, even if the shorter propagations are critically sampled, the final result is the same as a single propagation! It is the total propagation distance that is important; however, split-step simulations are applied in many situations for reasons such as propagating between a series of atmospheric turbulence phase screens.

Previously, it was noted that the reason the IR approach behaved better for the long propagation example is that it effectively suppresses source frequency content where the frequency chirp function is going bad. In fact, the IR approach is mainly introduced to give a quick and relatively easy way to approach longer propagation distances. But there are other ways to handle this issue. Researchers working with laser beam propagation simulations also apply window functions to either suppress the source spectrum or remove energy in the wings of the source field. This, combined with multi-step propagation, can give good results. This subject is covered in more detail by Schmidt in reference 2.

Suppose a simulation involves some fixed parameters in the source or observation planes such that a single side length and sample interval will not serve for modeling both planes. In this situation the ability to independently select the physical side lengths of the source and observation planes is helpful. The *two-step method* allows the source and observation plane side lengths to be different. This is described and analyzed in Appendix B. While it still suffers from some of the same sampling limitations described for the TF approach, it affords flexibility in the simulation design.

5.5 Fraunhofer Propagation

The expression for the Fraunhofer pattern is repeated here:

$$U_2(x_2, y_2) = \frac{\exp(jkz)}{j\lambda z} \exp\left[j\frac{k}{2z}(x_2^2 + y_2^2)\right]$$
$$\times \iint U_1(x_1, y_1) \exp\left[-j\frac{2\pi}{\lambda z}(x_2 x_1 + y_2 y_1)\right] dx_1 dy_1, \quad (5.16)$$

where, for coding purposes, the source plane variables are now indicated with the subscript 1 and the observation plane variables with subscript 2. When using the FFT to compute the Fraunhofer field, the source and observation plane side lengths are not generally the same. From Eq. (4.31),

$$\lambda z f_{X1} \rightarrow x_2, \quad (5.17)$$

and using Eq. (2.17), the observation plane side length and sample interval are found in terms of the source plane parameters

$$L_2 = \frac{\lambda z}{\Delta x_1}, \quad \text{and} \quad \Delta x_2 = \frac{\lambda z}{L_1}. \quad (5.18)$$

So, the observation plane coordinates are given as

$$x_2 = \left[\frac{-L_2}{2} : \Delta x_2 : \frac{L_2}{2} - \Delta x_2\right] = \lambda z \left[\frac{-1}{2\Delta x_1} : \frac{1}{L_1} : \frac{1}{2\Delta x_1} - \frac{1}{L_1}\right]. \quad (5.19)$$

If critical sampling is used ($\Delta x_1 = \lambda z/L_1$), then (5.18) indicates that the side lengths will be equal, $L_2 = L_1$. Otherwise, the side lengths are different. The function `propFF` that computes the Fraunhofer pattern follows:

```
1   function[u2,L2]=propFF(u1,L1,lambda,z);
2   % propagation - Fraunhofer pattern
3   % assumes uniform sampling
4   % u1 - source plane field
5   % L1 - source plane side length
6   % lambda - wavelength
7   % z - propagation distance
8   % L2 - observation plane side length
9   % u2 - observation plane field
10  %
11  [M,N]=size(u1);              %get input field array size
12  dx1=L1/M;                    %source sample interval
13  k=2*pi/lambda;               %wavenumber
14  %
15  L2=lambda*z/dx1;             %obs sidelength
16  dx2=lambda*z/L1;             %obs sample interval
17  x2=-L2/2:dx2:L2/2-dx2;       %obs coords
```

Propagation Simulation

```
18  [X2,Y2]=meshgrid(x2,x2);
19  %
20  c=1/(j*lambda*z)*exp(j*k/(2*z)*(X2.^2+Y2.^2));
21  u2=c.*ifftshift(fft2(fftshift(u1)))*dx1^2;
22  end
```

This function also outputs the side length of the observation plane L2 so it doesn't have to be computed again in the main script. To try this out, make the following changes in the sqr_beam routine:

```
w=0.011;                %source half width (m)

[u2,L2]=propFF(u1,L1,lambda,z);

dx2=L2/M;
x2=-L2/2:dx2:L2/2-dx2;  %obs  ords
y2=x2;

imagesc(x2,y2,nthroot(I2,3));%stretch image contrast
```

Use "Save as" and give this file a new name, sqr_beam_FF. The source half width w = 0.011 m with the propagation distance of z = 2000 m gives a Fresnel number of 0.12, which is reasonable for the Fraunhofer approximation. Running sqr_beam_FF gives the irradiance results of Fig. 5.7. Stretch the contrast of the irradiance pattern with the nthroot function to bring out the sidelobes.

The simulation result can be checked against the analytic Fraunhofer result. Take the Fourier transform of the source distribution:

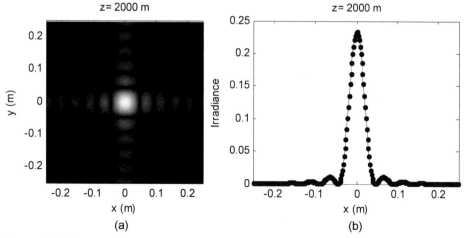

Figure 5.7 Fraunhofer irradiance (a) pattern and (b) x profile. Points in (b) are analytic values.

$$\Im\left\{\text{rect}\left(\frac{x_1}{2w}\right)\text{rect}\left(\frac{y_1}{2w}\right)\right\} = 4w^2 \text{sinc}(2wf_{X1})\text{sinc}(2wf_{Y1}). \quad (5.20)$$

Substitute $x_2/\lambda z$ for f_{X1} and $y_2/\lambda z$ for f_{Y1} and include the multipliers to get the Fraunhofer field:

$$U_2(x_2, y_2) = \frac{\exp(jkz)}{j\lambda z}\exp\left[j\frac{k}{2z}(x_2^2 + y_2^2)\right]$$
$$\times 4w^2 \text{sinc}\left(\frac{2w}{\lambda z}x_2\right)\text{sinc}\left(\frac{2w}{\lambda z}y_2\right).$$
(5.21)

The irradiance pattern is $I_2(x_2, y_2) = |U_2(x_2, y_2)|^2$, which yields

$$I_2(x_2, y_2) = \left(\frac{4w^2}{\lambda z}\right)^2 \text{sinc}^2\left(\frac{2w}{\lambda z}x_2\right)\text{sinc}^2\left(\frac{2w}{\lambda z}y_2\right). \quad (5.22)$$

Points for the analytic function are also plotted in Fig. 5.7(b). Now it is your turn: insert Eq. (5.22) into the script and see if you can get the plot shown in Fig. 5.7(b).

Usually, the irradiance is of interest when calculating the Fraunhofer pattern, so the complex exponentials out front disappear. In this case the only worry is the usual source plane sampling condition, $B_1 \leq 1/2\Delta x_1$. But, suppose the Fraunhofer *field* is of interest, including the chirp term. Based on Eq. (5.7), the chirp function $\exp[jk(2z)^{-1}(x_2^2 + y_2^2)]$ will be adequately sampled in the observation plane if $\Delta x_2 \leq \lambda z / L_2$, or equivalently, by applying Eq. (5.18) when the source plane sampling is

$$\Delta x_1 \geq \frac{\lambda z}{L_1}. \quad (5.23)$$

If Eq. (5.23) is not satisfied, the chirp phase will be aliased when tacked on the front of the transform. Furthermore, consider the Fresnel number, where $w^2 / \lambda z \ll 1$ in the Fraunhofer regime, and combine this with Eq. (5.23) to find

$$L_1 \gg \frac{w^2}{\Delta x_1}, \quad \text{or} \quad M \gg \frac{w^2}{\Delta x_1^2}, \quad (5.24)$$

which says the number of grid samples M needs to be much greater than the square of the number of samples across the source aperture radius $w^2/\Delta x_1^2$. This implies a large M. Fortunately, the Fraunhofer phase is not often required.

5.6 Coding Efficiency

Some aspects of the functions and scripts presented in this chapter (and throughout the book) are inefficient for reasons related to ease of use or presentation. For example, `meshgrid` is used in the `sqr_beam` code to define the sample coordinates, but then it is applied again in the propagation functions to redefine the coordinates. This makes the functions easier to use but it is redundant.

Speed and efficiency are not a big problem for the examples in this book, but they can be an important issue when running many iterations of a propagation code. To help tune-up your code, try out two coding tools provided by MATLAB: M-Lint and the Profiler.

M-Lint is an analyzer that checks the code in the Editor for possible problems. Get to M-Lint by going to the MATLAB Desktop toolbar. Select "Tools > Check Code with M-Lint." You will see a display of potential problems that M-Lint finds.

The Profiler tracks the execution time of the various statements and function calls in your code. It can help find problems and improve the efficiency of your code. On the Desktop toolbar, select "Tools > Open Profiler." The Profiler window will appear. The file to be profiled is in the menu box next to "Run this code:" Click on "Start Profiling." The code will execute, and a Profile Summary window will give a breakdown of the execution times.

5.7 Exercises

5.1 Assume a circular aperture with a radius of $w = 0.05$ m illuminated by a plane-wave, where $\lambda = 0.5$ μm. Assume a propagation distance of 1000 m and a simulation array size of 500 × 500 samples. Assume critical sampling for a Fresnel propagation.

(a) Find the side length L_1, sample interval Δx, and the Nyquist frequency F_N.

(b) Determine the source effective bandwidth B_1. Is $B_1 < F_N$? How many samples span the diameter of the circle function?

(c) Determine the Fresnel number. Is the propagation distance within the Fresnel region?

(d) Using the value for L_1 from (a), simulate Fresnel propagation for distances of 500, 1000, and 2000 m. Try both TF and IR simulations.

5.2 Return to the three apertures of Exercise 4.4. Simulate Fresnel propagation of the source fields in a 500×500 sample array with the following parameters:

Aperture (a), $L_1 = 2$ mm; $z = 0.5$, 1, and 5 cm.

Aperture (b), $L_1 = 2.5$ cm; $z = 0.5$, 2, and 5 m.

Aperture (c), $L_1 = 2.5$ cm; $z = 0.5$, 2, and 5 m.

What are the distances z that result in critical sampling?

5.3 A useful diagnostic for propagation simulations is to compute the power in the source and observation planes. Assuming no absorption or scatter of the light, which is true for the simulations presented in this book, the power (proportional to watts) should be conserved. In other words, the source and observation planes should contain the same optical power. If not, there may be a code error or a sampling problem. The power is the integrated irradiance, or

$$P = \iint I(x,y)\,dxdy. \tag{5.25}$$

For the sqr_beam example in this chapter:

(a) Add code to compute the power at the source and observation planes for the various propagations. Use the MATLAB function sum. Maybe two of these? What about dx and dy? You can remove the semicolon from the end of the line with the power calculation so that the value displays in the Command Window when the script is executed.

(b) Are there discrepancies in some of the sqr_beam example cases regarding the power in the source and observation planes?

5.4 Rayleigh–Sommerfeld diffraction: Fresnel diffraction involves a paraxial (small ray angle) assumption that limits the minimum propagation distance, whereas Rayleigh–Sommerfeld diffraction is essentially exact over all distances.

(a) Based on the Fresnel diffraction routines, write transfer function and impulse response propagators for *Rayleigh–Sommerfeld diffraction*.

(b) Test the routines with the sqr_beam example for the four distances (1000, 2000, 4000, and 20,000 m). Are there differences between the Fresnel and Rayleigh–Sommerfeld results?

(c) Calculate the Fresnel numbers for the `sqr_beam` examples. What can you say about applying Fresnel versus Rayleigh–Sommerfeld propagation in this case?

5.5 Gaussian Beam. Fourier methods are well suited for simulating laser beam propagation. Typically, a laser beam obeys the paraxial ray angle approximation, which is valid for the Fresnel expression. Also, the Gaussian function used to describe the beam profile is more forgiving in terms of sampling artifacts than a square or circular aperture beam of similar support. Laser textbooks define the irradiance distribution of a Gaussian laser beam (TEM$_{0,0}$ mode) propagating in the z direction as

$$I(x, y, z) = I_0 \left(\frac{w_0}{w(z)} \right)^2 \exp\left(-2 \frac{(x^2 + y^2)}{w(z)^2} \right), \quad (5.26)$$

where x, y are transverse spatial variables, I_0 is the source irradiance value at beam center ($x, y = 0$), w_0 = source beam e^{-2} radius (at $z = 0$), and $w(z)$ is the beam radius at distance z. The beam radius is given by

$$w(z) = w_0 \sqrt{1 + \left(\frac{z}{z_R} \right)^2}, \quad (5.27)$$

where z_R is the Rayleigh range defined by

$$z_R = \frac{\pi w_0^2}{\lambda}. \quad (5.28)$$

For the following questions, assume a source optical field ($z = 0$) given by

$$U_0(x, y) = A_0 \exp\left[-\frac{(x^2 + y^2)}{w_0^2} \right], \quad (5.29)$$

where $w_0 = 1$ mm, $\lambda = 0.633$ μm, and $A_0 = 1$ V/m. To be accurate, $I_0 = |A_0|^2/(2\eta)$ W/m^2 where $\eta = 377$ Ω in free space.

(a) Assume a side length of 15 mm and an array of 250 × 250 elements. Create the Gaussian beam of Eq. (5.29) in the source array, then simulate Fresnel propagation for distances of 1, 5, and 10 m. Compare irradiance results with the analytic result of Eq. (5.26).

(b) What is the propagation distance for critical sampling? Test the source bandwidth criterion for a 10-m propagation distance.

(c) Derive the Fraunhofer irradiance expression for the U_0 beam. Show that your result is consistent with the analytic expression in Eq. (5.26).

5.6 Test the concept expressed in Eq. (5.15). Use the TF propagator with the `sqr_beam` example to simulate a total propagation distance of $z = 20{,}000$ m, but make a split-step simulation where the propagator is called 10 times in succession. The propagation distance for each step is $\Delta z = 2000$ m (critical sampling). Compare the split-step result with a single TF propagation of 20,000 m. Are the results the same?

5.7 Return to the three apertures of Exercise 4.4. Simulate Fraunhofer propagation of the source fields in a 500×500 sample array with the following parameters:

Aperture (a): $L_1 = 2$ mm; $z = 5$ m.

Aperture (b): $L_1 = 2.5$ cm; $z = 50$ m.

Aperture (c): $L_1 = 2.5$ cm; $z = 50$ m.

Compare discrete and analytic results in an x-axis irradiance profile. (Note that there is no attempt in this exercise to model the Fraunhofer field such that the phase is adequately sampled.)

5.8 Consider a square aperture illuminated by a plane wave and assume the square function is sampled in a simulation such that at least 98% of the spectral power is available [see Eq. (2.11)]. Find a criterion for the number of linear samples M necessary for the simulation array in order to adequately sample the Fraunhofer field phase. The result should contain no variables—just a number.

5.9 Two-step propagator. Code up the two-step propagator function described in Appendix B. Test it with the `sqr_beam` example for the following cases:

(a) For $L_1 = 0.5$ m; $z = 2000$ m (critical sampling distance): Examine the field magnitude profiles for $L_2 = 0.2, 0.4, 0.5, 0.6$, and 1 m. The 0.5 m case should be identical to the TF propagation result. What are the apparent artifacts for the other distances?

(b) For $L_1 = 0.5$ m; $z = 20{,}000$ m: Does adjusting the L_2 size reduce artifacts in the observation plane result?

5.8 References

1. D. G. Voelz and M. C. Roggemann, "Digital simulation of scalar optical diffraction: Revisiting chirp function sampling criteria and consequences," *Appl. Opt.*, **48,** 6132–6142 (2009).
2. J. D. Schmidt, *Numerical Simulation of Optical Wave Propagation with Examples in MATLAB®*, SPIE Press, Bellingham, WA (2010). doi:[10.1117/3.866274].
3. J. W. Goodman, *Introduction to Fourier Optics*, 3rd ed., Roberts & Company, Greenwood Village, CO (2005).
4. M. C. Roggemann, and B. M. Welsh, *Imaging Through Turbulence*, CRC, Boca Raton, FL (1996).

Chapter 6
Transmittance Functions, Lenses, and Gratings

The beam sources implemented in Chapter 5 are for the most part simple apertures illuminated by a plane wave. They are modeled with real functions and, in effect, have a zero phase component. In this chapter functions are presented that create a more complicated field by altering the magnitude and/or phase of the field. The functions can be used to apply "tilt" or "focus" to a field, model the effect of a periodic structure, or model a lens. In general, these transmittance functions can be thought of as multiplying an incident field to create a desired effect; however, some represent well-known optical components such as a diffraction grating or a lens.

The functions discussed in this chapter provide considerable utility in their own right, but like the basic functions they can be combined to create more elaborate fields. As a matter of convenience these functions are described as part of the source, or as applied in the source plane. However, they can be applied in other planes; for example, the pupil of an imaging system, which is coming up in Chapter 7.

6.1 Tilt

An optical beam can be steered in a propagation simulation by applying a "tilt" to the beam wavefront. Suppose a tilt of angle of α is applied to a wavefront, as indicated in Fig. 6.1, where the dashed line represents the tilted wavefront of the beam and the arrow indicates the intended direction of propagation.

An expression for the dashed line in Fig. 6.1 is $z = -y \tan \alpha$. The intent is to convert this line to a phase front in the x–y plane at $z = 0$. This essentially requires replacing the position z with a phase quantity. As time progresses the wave moves in the positive z direction, but (as noted previously), the phase representation becomes more negative. This reverses the sign of the expression. The wavelength λ corresponds to 2π rad in the phase notation; so, using the wavenumber parameter $k = 2\pi/\lambda$, the phase function for producing the tilt is

$$\phi_Y(x, y) = ky \tan \alpha . \qquad (6.1)$$

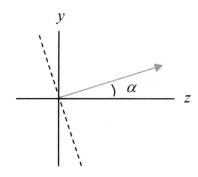

Figure 6.1 Wavefront tilt in the y–z plane.

More generally, to produce a tilt α relative to the z axis but in a radial direction defined by the angle θ in the x–y plane (see Fig. 6.2), one can use

$$\phi(x, y) = k(x\cos\theta + y\sin\theta)\tan\alpha , \qquad (6.2)$$

where $\theta = \tan^{-1}(y/x)$. The transmittance function is, therefore,

$$t_A(x, y) = \exp[jk(x\cos\theta + y\sin\theta)\tan\alpha]. \qquad (6.3)$$

Note that $\alpha = \tan^{-1}(r/z)$, where r is the radial distance in the observation plane from the origin to the beam aiming point (Fig. 6.2). The function `tilt` is listed here:

```
1  function[uout]=tilt(uin,L,lambda,alpha,theta)
2  % tilt phasefront
3  % uniform sampling assumed
4  % uin - input field
5  % L - side length
6  % lambda - wavelength
7  % alpha - tilt angle
8  % theta - rotation angle (x axis 0)
```

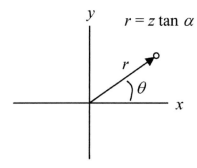

Figure 6.2 Beam-aiming point in the observation plane.

```
9   % uout - output field
10
11  [M,N]=size(uin);         %get input field array size
12  dx=L/M;                  %sample interval
13  k=2*pi/lambda;           %wavenumber
14
15  x=-L/2:dx:L/2-dx;        %coords
16  [X,Y]=meshgrid(x,x);
17
18  uout=uin.*exp(j*k*(X*cos(theta)+Y*sin(theta))...
19      *tan(alpha));        %apply tilt
20  end
```

In the expression for uout (line 18), typing three periods and hitting enter allows the equation to be continued on the next line. Test the tilt function by returning to the "sqr_beam" example from Section 5.3. Save a new version (sqr_beam_tilt) and insert the following before the propagation call:

```
deg=pi/180;
alpha=5.0e-5;  %rad
theta=45*deg;
[u1]=tilt(u1,L1,lambda,alpha,theta);
```

where Fig. 6.3 is the irradiance result after executing the example.

Sampling limitations also exist for this technique. As one might guess, if the tilt is large enough to translate the beam in the observation plane beyond the grid boundary, there will be trouble (see Exercise 6.1). To study this limitation, consider that tilt is a linear phase exponential applied to the source function U_1. Assume a single-axis tilt, and apply the shift theorem to the transform of the source field to get

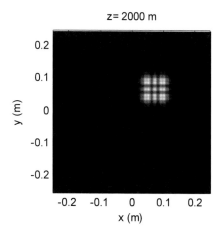

Figure 6.3 Irradiance pattern for the tilt example: $\alpha = 0.5$ μrad, $\theta = 45$ deg.

$$\Im\{U_1(x_1, y_1)\exp(jkx_1 \tan \alpha)\} = G_1\left(f_{X1} - \frac{\tan \alpha}{\lambda}, f_{Y1}\right), \qquad (6.4)$$

where $G_1 = \Im\{U_1\}$. If the source has a bandwidth of B_1, then considering the spectrum is essentially shifted, the combined effective bandwidth for the field leaving the source is

$$B_1^{+T} = B_1 + \frac{\tan \alpha}{\lambda}. \qquad (6.5)$$

This bandwidth can now be substituted for B_1 in the propagation criteria of Table 5.3. For example, if $\Delta x \leq \lambda z/L$, then the propagation criterion is $B_1^{+T} \leq 1/(2\Delta x)$. Using Eq. (6.5), and with some rearrangement, we get

$$\alpha \leq \lambda\left(\frac{1}{2\Delta x_1} - B_1\right), \qquad (6.6)$$

where α is assumed to be a small angle. Some comments about this criterion include the following:

(a) The result is approximate as the specific interaction of the source and propagator phase is not accounted for in Eq. (6.5) or (6.6). For example, one may cancel some of the effects of the other.

(b) If $\Delta x > \lambda z / L$ ("shorter distance"), a variety of artifacts can appear in the observation plane, such as "fringing" and asymmetries before the tilt angle α nears the bound.

(c) If $\Delta x < \lambda z / L$ ("longer distance"), replace $1/2\Delta x_1$ with $L_1/(2\lambda z)$ in Eq. (6.6).

In the `sqr_beam_tilt` example, $\lambda = 0.5 \times 10^{-6}$ m and $\Delta x_1 = 2 \times 10^{-3}$ m. Using $B_1 \approx 98$ cycles/m, which was calculated in Eq. (5.10), the tilt angle criterion result is $\alpha \leq 7.6 \times 10^{-5}$ rad. Try this tilt angle and see what happens. Set `theta=0` and `alpha=7.6e-5`. The resulting beam should appear close to the array edge. Check the magnitude profile where some "wrap-around" effects are apparent with energy from the beam entering the opposite side of the grid (periodic extension!). To split the pattern across the grid boundary, let $\alpha = \lambda/2\Delta x_1 = 1.25 \times 10^{-4}$.

Try some shorter and longer propagation distances in `sqr_beam_tilt` and observe how the beam is affected as the tilt increases. In general, it is a good idea to work with tilt angles that are well within the limit set by Eq. (6.6). In typical simulations the maximum available tilt angle is quite small, which is consistent with the paraxial nature of the Fresnel propagator.

6.2 Focus

Another useful operation is converging ("focusing") or diverging ("defocusing") an optical beam. A beam with a spherical wavefront, as shown in Fig. 6.4, will converge to the position z_f on the z axis. We can proceed in the same manner as was done for tilt to find the converging phase front in the x–y plane at $z = 0$. This is given by

$$\phi_S(x, y) = -k\sqrt{z_f^2 + x^2 + y^2} \;. \tag{6.7}$$

As x or y increases the phase values become more negative, which indicates the wavefront off the axis *leads* (in time) the on-axis wavefront (see Section 4.3). Thus, the negative sign in Eq. (6.7) corresponds to a *converging* wavefront, as illustrated in Fig. 6.4. A positive sign corresponds to a *diverging* wavefront. Borrowing from the discussion of the Fresnel diffraction in Section 4.4.2, the application of the binomial approximation gives a parabolic phase front that approximates the spherical phase front:

$$\phi(x, y) = -\frac{k}{2z_f}(x^2 + y^2). \tag{6.8}$$

The transmittance function for focus is, therefore,

$$t_A(x, y) = \exp\left[-j\frac{k}{2z_f}(x^2 + y^2)\right]. \tag{6.9}$$

This is a phase chirp function of the same form that appears in Eq. (4.23) for the Fresnel impulse response function h, although the exponent sign is negative.

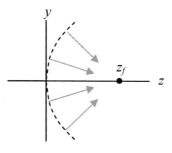

Figure 6.4 Converging wavefront.

A MATLAB function for applying focus follows:

```
1   function[uout]=focus(uin,L,lambda,zf)
2   % converging or diverging phase-front
3   % uniform sampling assumed
4   % uin - input field
5   % L - side length
6   % lambda - wavelength
7   % zf - focal distance (+ converge, - diverge)
8   % uout - output field
9
10  [M,N]=size(uin);         %get input field array size
11  dx=L/M;                  %sample interval
12  k=2*pi/lambda;           %wavenumber
13  %
14  x=-L/2:dx:L/2-dx;        %coords
15  [X,Y]=meshgrid(x,x);
16
17  uout=uin.*exp(-j*k/(2*zf)*(X.^2+Y.^2));  %apply focus
18  end
```

Try this on the sqr_beam example (new file sqr_beam_focus). Insert the following before the propagation call and run the script:

```
zf=2000;
[u1]=focus(u1,L1,lambda,zf);
```

The result is shown in Fig. 6.5. That looks like a pretty good focus! In this example the focal distance is the same as the propagation distance, so a small spot is expected. The pattern is, in fact, a scaled Fraunhofer pattern. (Check out Section 6.3.) Try some other focal distances—see what happens. Can you get the pattern to expand to fill the observation plane grid? A negative focus value puts the focal point in a virtual position behind the plane and causes a diverging wave (Fig. 6.6). Try it!

Multiplying a source field by Eq. (6.9) has the effect of increasing the source bandwidth. The increase in bandwidth is roughly the source support radius $D_1/2$ over $\lambda |z_f|$, so the combination effective bandwidth is approximately

$$B_1^{+F} \approx B_1 + \frac{D_1/2}{\lambda |z_f|}. \tag{6.10}$$

For $B_1^{+F} \leq 1/(2\Delta x)$, then with some rearranging, a bound is obtained for the focal distance:

Transmittance Functions, Lenses, and Gratings 95

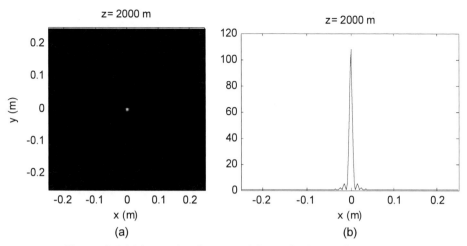

Figure 6.5 (a) Image irradiance and (b) profile for the focus example.

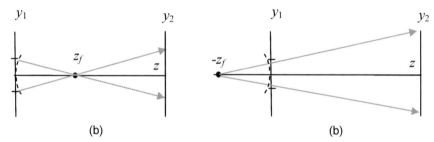

Figure 6.6 Geometrical ray diagram for (a) a converging wavefront and (b) a diverging wavefront at the source plane.

$$|z_f| \geq \frac{D_1/2}{\lambda}\left(\frac{1}{2\Delta x_1} - B_1\right)^{-1}. \quad (6.11)$$

As with the tilt angle criterion, Eq. (6.11) is approximate because the exact interaction of the source and propagator phase is not considered.

For the `sqr_beam_focus` example, $\lambda = 0.5 \times 10^{-6}$ m, $\Delta x_1 = 2 \times 10^{-3}$ m, $B_1 \approx$ 98 cycles/m and $D_1 = 0.102$ m, which leads to $|z_f| \geq 671$ m. Test this condition with the `sqr_beam_focus` code. When $z_f = -671$ m, the beam is diverging and clearly is pushing over the array boundaries in the observation plane. When $z_f = +671$ m, the pattern is smaller since the beam went through a focus before the observation plane (Fig. 6.6). The pattern looks reasonable; however, phase aliasing errors are just starting to creep in on the array edges. This can be seen in the unwrapped phase profile of the observation plane field.

6.3 Lens

A lens is an optical element that uses refraction to focus or diverge light. The transmittance function for an ideal, simple lens is given by[1]

$$t_A(x,y) = P(x,y)\exp\left[-j\frac{k}{2f}(x^2+y^2)\right], \qquad (6.12)$$

where f is known as the focal length and $P(x,y)$ is the pupil function. This is essentially the same complex exponential defined for focus with z_f replaced by f. A positive focal length produces a converging wavefront from a plane-wave input and a negative focal length produces a diverging wavefront. The pupil function accounts for the physical size of the lens—the opening available to collect light. For example, the most common lens pupil function is a circle

$$P(x,y) = \text{circ}\left(\frac{\sqrt{x^2+y^2}}{w_L}\right), \qquad (6.13)$$

where w_L is the radius of the lens aperture (not to be confused with the radius of curvature of the wavefront that exits the lens).

It is not always practical to implement the transmittance function of Eq. (6.12) in Fresnel propagation as was done for the focus example. This is because the focal length f is governed by the same criterion as z_f given in Eq. (6.11), and since f tends to be relatively short, a large number of samples are required. Assume a plane wave incident on the lens, which implies $B_1 \rightarrow 0$, then with some algebra the expression in Eq. (6.11) leads to

$$\frac{|f|}{D_L} \geq \frac{\Delta x}{\lambda}, \qquad (6.14)$$

where $D_L = 2w_L$. The ratio $|f|/D_L$ is known as the focal ratio, or the *f-number*, indicated by $f/\#$. With this substitution,

$$f/\# \geq \frac{\Delta x}{\lambda}. \qquad (6.15)$$

Practical lenses have $f/\#$s ranging from perhaps 2 to roughly 50 and diameters $2w_L$ of a few millimeters to maybe 100 mm. Take a typical value, $f/\# = 10$ (usually written $f/10$) and a diameter of 25 mm. Assume visible light $\lambda = 0.5 \times 10^{-6}$ m, then Eq. (6.15) yields

$$\Delta x \leq 5 \times 10^{-6} \text{ m}. \tag{6.16}$$

Furthermore, to implement Fresnel propagation, the array side length L needs to at least span the lens diameter. Thus, the linear number of samples in an array required to model this lens is

$$M \geq \frac{L}{\Delta x} = \frac{25 \times 10^{-3}}{5 \times 10^{-6}} = 5000, \tag{6.17}$$

which is a rather large array. Thus, modeling lenses directly with the Fresnel propagator is usually practical only for large $f/\#$s.

However, all is not lost for smaller $f/\#$s. If the field incident on the lens is $U_1(x_1, y_1)$, then the field exiting the lens is $U_1(x_1, y_1)t_A(x_1, y_1)$. Insert this into Eq. (4.25) for $U_1(x_1, y_1)$ and set $z = f$. The chirp functions in the integral cancel, and the result is

$$U_2(x_2, y_2) = \frac{\exp(jkf)}{j\lambda f} \exp\left[j\frac{k}{2f}(x_2^2 + y_2^2)\right]$$
$$\times \iint U_1(x_1, y_1) P(x_1, y_1) \exp\left[-j\frac{2\pi}{\lambda f}(x_2 x_1 + y_2 y_1)\right] dx_1 dy_1. \tag{6.18}$$

The expression in Eq. (6.18) shows the field at the *focal plane* of an ideal positive lens is simply the *Fraunhofer pattern* of the incident field with $z = f$.

Therefore, to find the field or irradiance pattern in the focal plane of a positive lens, including one with a small $f/\#$, the function "prop_FF" from the previous chapter can be applied replacing z with f.

Take the parameters from the $f/10$ lens example and assume U_1 is a unit amplitude plane wave. Select $M = 250$ and $L = 250$ mm. The irradiance pattern in Fig. 6.7 is generated for the focal plane using prop_FF (see Exercise 6.4). Try it! The focused irradiance pattern formed with an ideal circular-shaped lens, such as shown in Fig. 6.7, is known as the Airy pattern.

A special case of interest is when the source field is located in the front focal plane of a positive lens, a distance f from the lens (Fig. 6.8). For this arrangement the field at the focal plane is[1]

$$U_2(x_2, y_2) = \frac{\exp(jkf)}{j\lambda f}$$
$$\times \iint U_1(x_1, y_1) P(x_1 + x_2, y_1 + y_2) \exp\left[-j\frac{2\pi}{\lambda f}(x_2 x_1 + y_2 y_1)\right] dx_1 dy_1.$$
$$\tag{6.19}$$

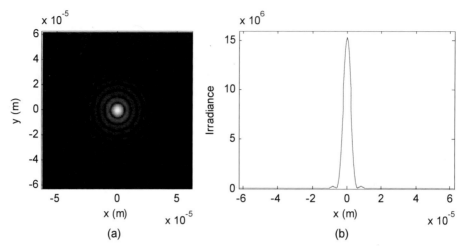

Figure 6.7 (a) Image irradiance and (b) x-axis profile for the lens Fraunhofer pattern. The fourth root is applied for (a). The large peak irradiance value in (b) is because all of the power in the unit amplitude field incident on the lens is being focused to a very small area.

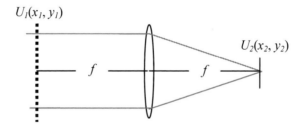

Figure 6.8 "Fourier transform" lens arrangement.

The chirp phase factor out front is now gone, so the focal plane field is a scaled *Fourier transform* of the input field. The arguments in the pupil function account for *vignetting*, which is a loss of light for off-axis points in the input field due to the finite pupil size. The effect of vignetting is reduced if the lens pupil is oversized compared to the support of the input field.

6.4 Gratings and Periodic Functions

A grating is an optical component that has a spatially periodic structure. Incident light diffracts either in transmission or reflection from the structure, and the colors (wavelength components) of the light become spatially separated some distance from the grating. Gratings are commonly used in spectrometers for examining the wavelength spectrum of an optical signal or in spectrophotometers that measure the spectral characteristics of an optical component. The diffraction pattern from a grating is usually observed in the Fraunhofer region.

6.4.1 Cosine magnitude example

A conventional grating has grooves cut into its surface that impart a magnitude and/or phase disturbance to the incident wave. To demonstrate modeling of periodic functions like those in gratings, start with an amplitude transmittance function given by

$$t_A(x,y) = \frac{1}{2}\left[1 - \cos\left(2\pi \frac{x}{P}\right)\right] \text{rect}\left(\frac{x}{D_1}\right) \text{rect}\left(\frac{y}{D_1}\right). \quad (6.20)$$

Figure 6.9 illustrates a 1D profile of this grating. In Eq. (6.20) the grating is defined within the 2D area $D_1 \times D_1$. The cosine pattern is only a function of x and has a period P. Typically, $P \ll D_1$. When illuminated by a unit amplitude plane wave, the source field is $U_1(x_1, y_1) = t_A(x_1, y_1)$. The Fraunhofer pattern is created using a lens (or mirror) of focal length f. The grating is simulated in the script "grating_cos," where the first part of this code is shown here:

```
1   % grating_cos diffraction grating example
2
3   lambda=0.5e-6;    %wavelength
4   f=0.5;            %focal distance
5   P=1e-4;           %grating period
6   D1=1.02e-3;       %grating side length
7
8   L1=1e-2;          %array side length
9   M=500;            %# samples
10  dx1=L1/M;
11  x1=-L1/2:dx1:L1/2-dx1; %source coords
12  [X1,Y1]=meshgrid(x1,x1);
13
14  % Grating field and irradiance
15  u1=1/2*(1-cos(2*pi*X1/P)).*rect(X1/D1).*rect(Y1/D1);
16
17  % Fraunhofer pattern
18  [u2,L2]=propFF(u1,L1,lambda,f);
19  dx2=L2/M;
20  x2=-L2/2:dx2:L2/2-dx2; y2=x2;%obs coords
21  I2=abs(u2).^2;
```

It is critical that the periodic function be sampled adequately. The number of samples that span the periodic function (the length P) is

$$\text{\# samples across } P = M \frac{P}{L_1}. \quad (6.21)$$

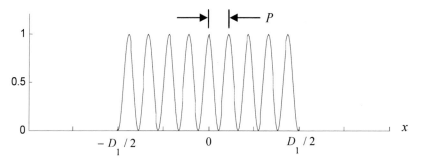

Figure 6.9 Cosine grating profile.

In the case of grating_cos, $MP/L_1 = 5$ indicates that five samples span each cosine cycle, which is okay. At least two are required to satisfy the sampling theorem. The values for L1, D1, P, and M were also selected in grating_cos to provide a clear display of the Fraunhofer pattern. Furthermore, the value for D1 provides an odd number of samples (D1/dx1 = 51) across the rect functions, which is consistent with our rect function definition. The lens focal length was arbitrarily chosen as f = 0.5 m. Figure 6.10 shows irradiance images of the source plane (I1) and the observation plane (I2) and an *x*-axis profile in the observation plane. The central feature is known as the zero order and the two side features are the −1 and +1 "first-order" peaks.

To make sure the simulation is working properly, the results can be compared with the analytic expression for the Fraunhofer pattern. First, the Fourier transform of the source field is required:

$$\Im\{U_1(x_1,y_1)\} = \frac{1}{2}\left[\delta(f_{X1},f_{Y1}) - \frac{1}{2}\delta\left(f_{X1}+\frac{1}{P},f_{Y1}\right) - \frac{1}{2}\delta\left(f_{X1}-\frac{1}{P},f_{Y1}\right)\right]$$
$$\otimes D_1^2 \operatorname{sinc}(D_1 f_{X1})\operatorname{sinc}(D_1 f_{Y1}). \quad (6.22)$$

Then perform the convolution:

$$\Im\{U_1(x_1,y_1)\} = \frac{D_1^2}{2}\operatorname{sinc}(D_1 f_{Y1})$$
$$\times \left\{\operatorname{sinc}(D_1 f_{X1}) - \frac{1}{2}\operatorname{sinc}\left[D_1\left(f_{X1}+\frac{1}{P}\right)\right] - \frac{1}{2}\operatorname{sinc}\left[D_1\left(f_{X1}-\frac{1}{P}\right)\right]\right\}. \quad (6.23)$$

Substitute $x_2/\lambda f \to f_{X1}$ and $y_2/\lambda f \to f_{Y1}$ and apply the multipliers to get the Fraunhofer field:

Transmittance Functions, Lenses, and Gratings 101

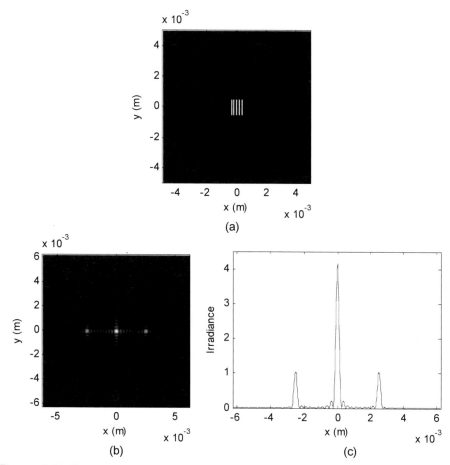

Figure 6.10 Results for grating_cos: (a) source irradiance; (b) Fraunhofer irradiance (contrast enhanced with third root); and (c) x-axis profile of Fraunhofer irradiance. [The display in (a) does not resolve the grating periodic features.]

$$U_2(x_2, y_2) = \frac{\exp(jkf)}{j\lambda f} \exp\left[j\frac{k}{2f}\left(x_2^2 + y_2^2\right)\right]\frac{D_1^2}{2}\text{sinc}\left(\frac{D_1}{\lambda f}y_2\right)$$

$$\cdot \left\{\text{sinc}\left(\frac{D_1}{\lambda f}x_2\right) - \frac{1}{2}\text{sinc}\left[\frac{D_1}{\lambda f}\left(x_2 + \frac{\lambda f}{P}\right)\right] - \frac{1}{2}\text{sinc}\left[\frac{D_1}{\lambda f}\left(x_2 - \frac{\lambda f}{P}\right)\right]\right\}.$$

(6.24)

The irradiance is the squared magnitude of Eq. (6.24). The following script portion evaluates the analytic irradiance result:

```
%analytic
[X2,Y2]=meshgrid(x2,y2);
lf=lambda*f;
u2a=(1/lf)*D1^2/2*sinc(D1/lf*Y2)...
    .*(sinc(D1/lf*X2)-1/2*sinc(D1/lf*(X2+lf/P))...
```

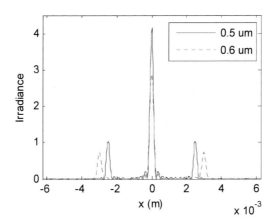

Figure 6.11 Irradiance profiles for grating_cos for λ = 0.5 μm and 0.6 μm.

```
    -1/2*sinc(D1/lf*(X2-lf/P)));
I2a=abs(u2a).^2;
```

The front complex exponential terms for the Fraunhofer pattern were not coded up since only the irradiance is being examined—but don't forget $1/\lambda f$. Try this and see if the discrete and analytic plots come out the same.

The main application for a grating is wavelength separation. Figure 6.11 shows superimposed curves for λ = 0.5 μm and λ = 0.6 μm. The first-order peaks are clearly separated. To produce these curves, run the code for one wavelength and in the command window store `I2` and `x2` in temporary arrays (for example, `I2p5=I2; x2p5=x2;`). Rerun the code for the other wavelength and execute the plot function for both curves under the Command Window using `plot(x2p5,I2p5(M/2+1,:),x2,I2(M/2+1,:))`. The x2 scale is different for each wavelength—don't use the same x2 vectors for the two curves! "Illuminating" a larger grating area narrows the orders and results in better spectral resolution. You can test this by making D_1 larger.

6.4.2 Square-wave magnitude example

Another example of a grating is illustrated in Fig. 6.12. The transmittance function is given by

$$t_A(x,y) = \left[\text{rect}\left(\frac{x}{P/2}\right) * \frac{1}{P}\text{comb}\left(\frac{x}{P}\right)\right]\text{rect}\left(\frac{x}{D_1}\right)\text{rect}\left(\frac{y}{D_1}\right), \quad (6.25)$$

where P is the period of a square-wave amplitude pattern that stretches in the x direction and lies within the 2D area of $D_1 \times D_1$. This grating is simulated in the following script "`grating_sqr`" (only a portion is shown):

Transmittance Functions, Lenses, and Gratings

```
1   % grating_sqr diffraction grating example
2
3   lambda=0.5e-6;    %wavelength
4   f=0.5;            %focal distance
5   P=1e-4;           %grating period
6   D1=1e-3;          %grating side length
7
8   L1=1e-2;          %array side length
9   M=1000;           %# samples
10  dx1=L1/M;
11  x1=-L1/2:dx1:L1/2-dx1; %source coords
12  [X1,Y1]=meshgrid(x1,x1);
13
14  % construct grating field
15  fc=fft(fftshift(ucomb(x1/P)));
16  fr=fft(fftshift(rect(x1/(P/2))));
17  ux=ifftshift(ifft(fc.*fr));  %1D conv rect & comb
18  u1=repmat(ux,M,1);           %replicate to 2D
19  u1=u1.*rect(X1/D1).*rect(Y1/D1); %set size
```

... (then apply the Fraunhofer propagator).

In this example the ucomb function is used to create a 1D periodic sequence of unit sample delta functions (defined in Appendix C). This delta is defined as a unit value at the coordinate of interest. The ucomb function truncates the input values at the sixth decimal position, so small round-off error will not cause problems in placing the unit samples.

For the ucomb function to work properly, the vector coordinates must be such that a sample is found at each position where the delta function is needed. Thus, P/dx1 needs to be an integer. In this case, dx1=1×10^{-5} m. Therefore the unit values appear in the vector fc for every 10 samples (P/dx1 = 10). An odd number of samples is arranged across each of the periodic rect functions (P/2 = 5) to be consistent with the rect function definition.

The approach for grating_sqr is to perform a 1D convolution (fft rather

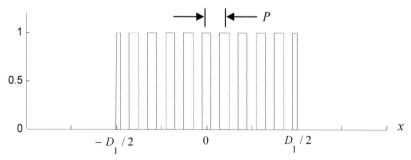

Figure 6.12 Square-wave grating profile.

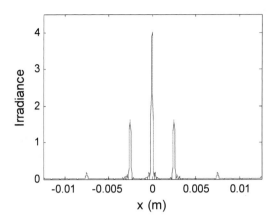

Figure 6.13 Fraunhofer irradiance profile for the grating modeled in grating_sqr.

than fft2) with the ucomb and rect sequences. A scaling multiplier is not needed on the convolution result because (a) the ifft correctly scales for one fft (for example, fr) and (b) each unit sample delta behaves as though it has an area of 1/dx1, which cancels the dx1 that arises from the second fft. The repmat function is a quick way to fill the rows of the 2D array u1 with the vector ux.

The field created by the grating is propagated using propFF, and the profile for I2 is shown in Fig. 6.13. It is similar to the cosine grating but with some additional higher-order peaks. Again, analytic theory can be used to verify the simulation. To find the Fraunhofer pattern, take the Fourier transform of the field and simplify as follows:

$$\Im\{U_1(x_1,y_1)\} = \left[\frac{P}{2}\text{sinc}\left(\frac{P}{2}f_{X1}\right)\text{comb}(Pf_{X1})\delta(f_{Y1})\right]$$
$$\otimes\left[D_1\text{sinc}(D_1f_{X1})D_1\text{sinc}(D_1f_{Y1})\right]$$
$$= \left[\frac{P}{2}\text{sinc}\left(\frac{P}{2}f_{X1}\right)\frac{1}{P}\sum_{n=-\infty}^{\infty}\delta\left(f_{X1}-\frac{n}{P}\right)\delta(f_{Y1})\right]$$
$$\otimes\left[D_1\text{sinc}(D_1f_{X1})D_1\text{sinc}(D_1f_{Y1})\right]$$
$$= \frac{D_1^2}{2}\text{sinc}(D_1f_{Y1})\sum_{n=-\infty}^{\infty}\text{sinc}\left(\frac{n}{2}\right)\text{sinc}\left[D_1\left(f_{X1}-\frac{n}{P}\right)\right]. \quad (6.26)$$

The Fraunhofer field is given by

Transmittance Functions, Lenses, and Gratings

$$U_2(x_2,y_2) = \frac{\exp(jkf)}{j\lambda f}\exp\left[j\frac{k}{2f}(x_2^2+y_2^2)\right]\frac{D_1^2}{2}\mathrm{sinc}\left(\frac{D_1}{\lambda f}y_2\right)$$
$$\cdot \sum_{n=-\infty}^{\infty}\mathrm{sinc}\left(\frac{n}{2}\right)\mathrm{sinc}\left[\frac{D_1}{\lambda f}\left(x_2-n\frac{\lambda f}{P}\right)\right].$$ (6.27)

Below is part of the code to generate a 1D irradiance slice for Eq. (6.27). A `for` loop is used for the summation and 11 terms are computed (n = −5 to 5). More terms do not make a noticeable difference in the result:

```
lf=lambda*f;
u2a=zeros(1,M);
for n=-5:5
    ut=sinc(n/2)*sinc(D1/lf*(x2-n*lf/P));
    u2a=u2a+ut;
end
u2a=D1^2/(2*lf)*u2a;
```

Implement this code at the end of `sqr_grating`. Compare the results with the numerical simulation. Can you get them to match?

6.4.3 One-dimensional model

The gratings examined thus far have no y directional periodic dependence. When this is the case, a 1D model (where $y = 0$) can be used to analyze the most critical part of the grating response, which is along the x axis. To do this, the y-dependent terms are removed, `meshgrid` is no longer required, and the function `repmat` is not required. Also, a 1D Fraunhofer calculation is required. The advantage of 1D modeling is that larger vectors can be used; so, more overall width and cycles across the grating can be modeled.

Return to the `grating_sqr` example and make the changes listed above for a 1D result. Call this new script "`grating_sqr1D`." Let M=2200, but leave all of the physical parameter definitions the same as before. Be sure to remove `rect(Y1/D1)` from line 19. The 2D function `propFF` is removed and the 1D Fraunhofer pattern can be calculated using

```
lf=lambda*f;
u2=sqrt(1/lf)*ifftshift(fft(fftshift(u1)))*dx1;
L2=lf/dx1;
```

Figure 6.14 shows profiles of the 1D Fraunhofer pattern. The greater number of samples results in an increase of the side length of the Fraunhofer pattern. The expanded view in Fig. 6.11(b) shows the relative size and position of the orders are the same as in the original result in Fig. 6.10. The overall magnitudes are

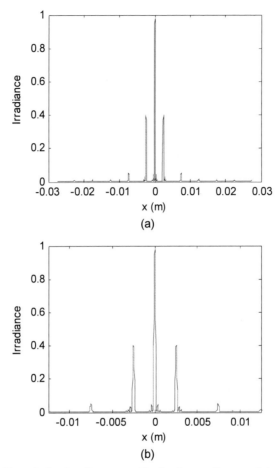

Figure 6.14 1D Fraunhofer irradiance profile for the grating modeled in grating_sqr: (a) full vector view; (b) expanded view.

different because the 1D case does not account for power associated with the second dimension. More samples across the periodic features in the 1D model produce fewer artifacts (small ripples) in the result compared to the 2D result.

6.4.4 Periodic model

How about modeling a grating that is essentially infinite in extent? Although not a practical device, it is not uncommon for a beam of light incident on a grating to cover hundreds or thousands of the periodic cycles—more than anything modeled here so far. But, more importantly, for an infinite-support grating the diffracted orders for an infinite (analytic) grating are delta functions and the multipliers for these delta functions indicate the relative amount of optical power that is directed to each order. For optical spectral analysis, high diffraction efficiency into the first order is usually desired.

Suppose the grating structure fills the array ($D_1 = L_1$) and the periodic features are carefully arranged such that an *integer number of periods P* exactly spans the array, or

$$\frac{L_1}{P} = \text{integer} .\qquad(6.28)$$

This case is illustrated in Fig. 6.15. The discrete Fourier transform, applied to obtain the Fraunhofer pattern, produces a result that is associated with repeating copies of the input array (periodic extension). Since the source is arranged to be perfectly continuous at the array boundaries, the result is the transform of an infinite periodic structure.

Return to Section 6.4.1 and the `grating_cos` example. Set `D1=1e-2`, which is the value of the side length `L1`. Since $L_1/P = 100$, there are exactly 100 cycles of the cosine function across the array. Executing the script yields sample delta functions at the diffractive order positions. To find the fraction of optical power in each order, compute the optical power at each sample in the observation

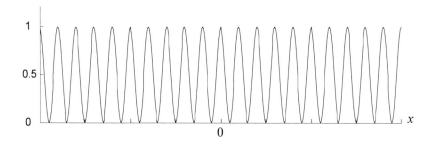

Figure 6.15 Cosine grating profile with an integer number of periods occupying the vector.

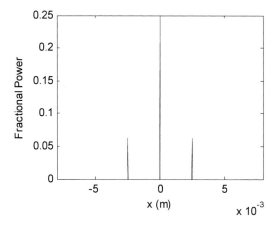

Figure 6.16 Fractional power result for a periodic model of the grating_cos example.

plane (irradiance integrated over the sample interval area) and divide by the total power transmitted by the grating (source field magnitude squared, integrated over the grating area). For `grating_cos` this operation looks like

```
Fpwr=I2*dx2*dx2/(L1^2);
```

where a unit source field magnitude is assumed. A display of `Fpwr` is shown in Fig. 6.16 for the `grating_cos` example; 25% of the power is in the zero order, and 6.25% is found in each of the ± first orders.

6.5 Exercises

6.1 The diagram in Fig. 6.17 illustrates the propagation of a field at an angle α relative to the optical axis. The field support D_2 just reaches to the edge of observation plane side length, a distance $L/2$ from the axis. Use this arrangement to derive the same tilt criterion as defined in Eq. (6.6). Assume critical sampling and Fraunhofer propagation. For Fraunhofer propagation, $D_2 = 2B_1 \lambda z$.

6.2 Demonstrate both tilt and focus simultaneously for the `sqr_beam` example. Use the TF approach with $z = 2000$ m. Assume $\alpha = 5.0 \times 10^{-5}$ rad, $\theta = 45$ deg for tilt and $z_f = 4000$ m for focus.

6.3 Among other applications, cylindrical lenses are commonly used to change an oblong-shaped laser diode beam into a more circular beam.

(a) Create a function to produce a *cylindrical focus* (focus in only one transverse axis).

(b) Demonstrate the cylindrical focus with the `sqr_beam` example, $z_f = 2000$ m.

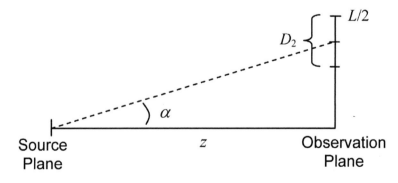

Figure 6.17 Diagram for propagation at an angle α.

(c) Is it generally possible to describe the irradiance in the focal plane of a cylindrical lens in terms of a Fraunhofer pattern as was done in Eq. (6.18) or (6.19) for a spherical lens?

6.4 Airy pattern:

(a) Generate the Airy pattern shown in Fig. 6.7 for an ideal $f/10$ lens ($f = 250$ mm and diameter $2w_L = 25$ mm). Use $M = 250$, $L = 250$ mm, and $\lambda = 0.5$ µm. Derive the analytic result and compare with the simulation result in a profile plot.

(b) Demonstrate the effect in the focal plane of reducing the diameter of the lens pupil.

6.5 Zone plate: Zone plates can be used in parts of the electromagnetic spectrum where glass or other materials are not transparent; but, for what purpose? Let's find out. Consider the transmittance function

$$t_A(r) = \frac{1}{2}\left[1 + \cos\left(\frac{k}{2f}r^2\right)\right]\mathrm{circ}\left(\frac{r}{w}\right), \qquad (6.29)$$

where $r = \sqrt{x^2 + y^2}$. The transmittance of this plate is illustrated in Fig. 6.18.

(a) Rewrite Eq. (6.29) using complex phasor notation. If a plane wave illuminates this plate, how do you expect the transmitted field to behave?

(b) Derive a sampling criterion for rendering this plate in a discrete array (check out Appendix A). Create a Fresnel propagation simulation for this plate in the source plane. Assume the following: unit amplitude

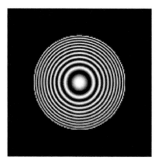

Figure 6.18 Zone plate transmittance illustration.

plane-wave illumination, $L_1 = 50$ mm, $M = 500$, $\lambda = 0.5$ μm, $w = 6.25$ mm, and $f = 10$ m. The choice of an extremely large value of f relative to the plate radius is necessary for sampling, and also to provide a magnified pattern at the observation plane.

(c) Show whether the sampling criterion for the plate is satisfied.

(d) Examine the propagation sampling for $z = 10$ m. Should the transfer function or impulse response approach be used?

(e) Simulate the irradiance patterns for propagation distances of 5, 8, 10, and 12 m. Display the patterns and profiles.

Another type of zone plate is given by

$$t_A(r) = \frac{1}{2}\left\{1 + \mathrm{sign}\left[\cos\left(\frac{k}{2f}r^2\right)\right]\right\}\mathrm{circ}\left(\frac{r}{w}\right), \qquad (6.30)$$

where the sign function is

$$\mathrm{sign}(x) = \begin{cases} -1, & x < 0. \\ 0, & x = 0. \\ 1, & x > 0. \end{cases} \qquad (6.31)$$

This plate has a binary transmittance (either 1 or 0) and can be made with rings of suspended opaque material.

(f) Simulate this plate and repeat part (e) using the same parameters as for the first plate (…you might check to see if MATLAB has a "sign" function…).

(g) Compare the irradiance profiles for the two plates at the focal plane ($z = 10$ m). Which is more efficient?

(h) Compare the zone plate profiles with the focal plane irradiance profile for a positive lens of the same characteristics.

6.6 How critical is the sampling arrangement for periodic functions? Do some testing! Change the number of samples in the `grating_sqr` example, for example, by +2. What happens? Try some other values (make sure M is still even so that other numerical issues are not also happening). What is the next value above $M = 1000$, where the periodic functions are again sampled appropriately?

6.7 Phase gratings: Unlike an amplitude grating, an ideal phase grating does not attenuate the incident light. Diffraction occurs because of periodic optical path length changes across the grating. A reflection grating can be modeled in the computer as a phase grating. Alter the `grating_cos` or grating_sqr code to compute the Fraunhofer pattern for the transmittance functions that follow. Adjust the factor m; for example, π, $\pi/2$, and $\pi/4$, and notice the effect on the Fraunhofer pattern.

(a) $t_A(x,y) = \exp\left[jm\cos\left(2\pi\frac{x}{P}\right)\right]\text{rect}\left(\frac{x}{D_1}\right)\text{rect}\left(\frac{y}{D_1}\right),$

(b) $t_A(x,y) = \exp\left[jm\,\text{rect}\left(\frac{x}{P/2}\right) * \frac{1}{P}\text{comb}\left(\frac{x}{P}\right)\right]\text{rect}\left(\frac{x}{D_1}\right)\text{rect}\left(\frac{y}{D_1}\right),$

(c) $t_A(x,y) = \exp\left[jm\Lambda\left(\frac{x}{P/2}\right) * \frac{1}{P}\text{comb}\left(\frac{x}{P}\right)\right]\text{rect}\left(\frac{x}{D_1}\right)\text{rect}\left(\frac{y}{D_1}\right).$

6.8 Consider the following grating transmittance function:

$$t_A(x,y) = \left[\Lambda\left(\frac{x}{P/2}\right) * \frac{1}{P}\text{comb}\left(\frac{x}{P}\right)\right]\text{rect}\left(\frac{x}{D_1}\right). \qquad (6.32)$$

(a) Implement this grating in a 1D simulation and compute the Fraunhofer pattern. Assume a grating frequency of 500 lines/mm [$P = (1/500) \times 10^{-3}$) m], $D_1 = 0.1$ mm, $\lambda = 0.5$ μm, and $f = 0.5$ m. Choose the vector size and other sampling parameters. [Hint: An even number of samples—for example, 8—across each triangle function works well in this case.]

(b) Derive a theoretical result for the Fraunhofer pattern for this grating (in 1D). Compare this result with the numerical result of part (a).

(c) Implement the grating in an infinite periodic model. Plot the Fraunhofer pattern profile. Plot the power percentage result.

6.6 References

1. J. W. Goodman, *Introduction to Fourier Optics*, 3rd Ed., Roberts & Company, Greenwood Village, CO (2005).

Chapter 7
Imaging and Diffraction-Limited Imaging Simulation

Imaging is about reproducing the field, or more often the irradiance pattern of an object or scene, at an image plane. Geometrical optics, where optical rays are assumed to travel in rectilinear fashion without diffraction, is used extensively in lens and optical system design. Geometrical optics provides useful relationships between the object and image locations and sizes and is also applied in the analysis of the pupils of an imaging system. A proficient approach for image modeling draws on both geometrical optics and diffraction theory. This chapter begins with a review of geometrical imaging concepts and relationships that are helpful for the imaging simulations that follow.

7.1 Geometrical Imaging Concepts

Not all optical systems form images. For example, a beam expander increases the size of a laser beam but doesn't image. However, our concern is with imaging, and in order to form a real image, light from an arbitrary object point must be collected and focused at the image plane. For the imaging situation shown in Fig. 7.1, the lens law (Gaussian form) describes the relationship needed under the paraxial condition (small ray angles relative to the optical axis) for "best focus" imaging:

$$\frac{1}{z_1} + \frac{1}{z_2} = \frac{1}{f}. \tag{7.1}$$

Here, f is the lens (or lens system) focal length, z_1 is the distance along the optical axis from the object to the *front principal plane* of the lens, and z_2 is the distance from the *back principal plane* to the image location. Principal planes are a virtual concept for geometrical lens analysis. They are normal to the optical axis. A ray incident on the front principal plane at some height from the optical axis will exit the back principal plane at the same height. In other words, principal planes are planes of unit magnification. For a "thin" lens, the front and back principal planes are co-located in the plane with the vanishingly thin lens. For a real (thick)

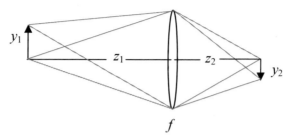

Figure 7.1 Geometrical imaging with a thin positive lens of focal length f. A cone of rays from the base or tip of the object are collected by the lens and directed to the corresponding image points.

lens, the principal planes are, typically, in the vicinity of the lens but not co-located.

To form a real image, z_1 and z_2 are positive and the lens focal length f is positive. A "positive" lens (positive-valued f) converges light rays, whereas a "negative" lens (negative-valued f) diverges rays. Practical imaging systems often use combinations of lenses to control aberrations or for packaging reasons, but imaging still requires a positive focal length for the combined lens group.

The ratio of the image height y_2 to the object height y_1 is known as the *transverse magnification* M_t, which for a single lens system is given by

$$M_t = \frac{y_2}{y_1} = -\frac{z_2}{z_1}. \qquad (7.2)$$

The minus sign indicates an inverted image (y_2 in Fig. 7.1 takes a negative value).

An imaging system is also characterized by its *pupils*. Pupils are virtual apertures that indicate the "opening" available to collect light from the object (*entrance pupil*) and the "opening" from which the collected light exits on its way to form an image (*exit pupil*). The pupils are images of the physical element in the system, known as the aperture *stop*, which limits the collection of light. The lens is the stop for the system in Fig. 7.1, but for a different system it might be some other physical aperture. For this discussion, the importance of the stop is that it sets the fundamental diffractive effects in the image—it's the thing that "cookie cutters" the incoming electric field. There are other system issues, such as aberrations, that further disrupt the image, but the diffractive effects due to the stop represent the fundamental performance limit of an imaging system. The stop and other system effects can all be incorporated in the pupils, so this concept is utilized for diffraction analysis.

Figure 7.2 illustrates that the physical elements of a system (lenses, mirrors, iris, etc.) can be reduced to entrance pupil (EP) and exit pupil (XP) models. The distance from an object point on the optical axis to the EP is z_{EP} and the distance from the XP to the axial image point is z_{XP}. The entrance pupil diameter is D_{EP} and the exit pupil diameter is D_{XP}. Although Fig. 7.2 shows

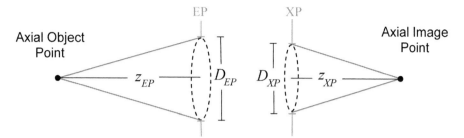

Figure 7.2 Entrance pupil (EP) and exit pupil (XP) model of an imaging system.

the EP nearest the object, in fact, the order/location of the pupils and sizes depend on the optical system being considered.

In a single thin lens system (Fig. 7.1), the principal planes, pupils, and lens all lie in the same plane, so $z_1 = z_{EP}$ and $z_2 = z_{XP}$, and also $D_{EP} = D_{XP}$ = lens diameter. This is a useful case to fall back on when thinking about the examples in this chapter. But, to provide some food for thought, refer to Fig. 7.3, where an object is being imaged and an iris located behind the lens limits the collection of light. The iris is the stop, and the exit pupil (a virtual aperture) is co-located with the iris and has the same diameter as the iris. The EP is found by imaging the stop (iris) "back through" the lens, which in this case produces the virtual aperture (EP), as shown in Fig. 7.3. In this case the pupil and principal plane distances are different.

One parameter that appears in expressions for the amount of light collected, aberration coefficients, spatial frequency content of an image, and other characteristics of an imaging system is the *f-number* ($f/\#$). This parameter was briefly introduced in Chapter 6 in the discussion of lenses. There are several definitions for the $f/\#$ that are applicable in different situations, but for this discussion the most useful form is the *paraxial working f/#* given by

$$f/\# = \frac{z_{XP}}{D_{XP}}. \tag{7.3}$$

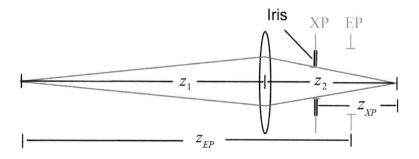

Figure 7.3 Entrance pupil (EP) and exit pupil (XP) for a single lens system imaging an object where the iris is the stop.

This is a parametrization of the cone of rays that travel from the XP to the axial image point (Fig. 7.2).

A summary of the key points of the geometrical optics discussion is as follows:

(a) The principal plane distances (z_1, z_2) define the transverse magnification of the image.

(b) The system aperture stop leads to the fundamental diffractive effects in the image.

(c) The pupil sizes (D_{EP}, D_{XP}) and distances (z_{EP}, z_{XP}) are incorporated in the diffraction analysis of the system.

(d) For a thin lens imaging system, $z_1 = z_{EP}$, $z_2 = z_{XP}$, and $D_{EP} = D_{XP} =$ lens diameter.

We refer the reader to other references for further discussions of principal planes, pupils, and geometrical optical imaging.[1-3]

7.2 Coherent Imaging

7.2.1 Coherent imaging theory

Now for some Fourier optics. The general imaging arrangement considered is shown in Fig. 7.4. Imaging with coherent illumination, such as with a coherent laser, is described in its simplest form as a convolution operation involving the optical *field*. The process is expressed by[4]

$$U_i(u,v) = h(u,v) \otimes U_g(u,v), \quad (7.4)$$

where u and v are the image plane spatial coordinates, U_i is the field at the image plane, and h is the coherent impulse response for the imaging system. U_g is the ideal geometrical-optics predicted image field, which is a scaled copy of the object field $U_o(x, y)$,

$$U_g(u,v) = \frac{1}{|M_t|} U_o\left(\frac{u}{M_t}, \frac{v}{M_t}\right). \quad (7.5)$$

Note that if M_t is negative, as in Eq. (7.2), then the resulting image will appear inverted relative to the object. In Eq. (7.4) the ideal geometrical field is "blurred" through the convolution with the impulse function. In the frequency domain the corresponding spectra for Eq. (7.4) are related by

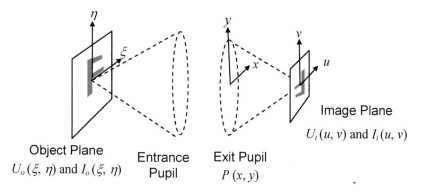

Figure 7.4 Imaging simulation coordinate definitions.

$$G_i(f_U, f_V) = H(f_U, f_V) G_g(f_U, f_V), \qquad (7.6)$$

where H is the *coherent image transfer function* (or amplitude transfer function) and is defined as[4]

$$H(f_U, f_V) = P(-\lambda z_{XP} f_U, -\lambda z_{XP} f_V), \qquad (7.7)$$

where P is the pupil function of the system, in this case the exit pupil since we are dealing with the image plane. Thus, the coherent transfer function takes on the attributes of the XP. A few comments about Eq. (7.7) follow:

(a) The negative signs in the pupil arguments give a scaled, inverted pupil function. This inversion is associated with the inversion of the ideal geometrical image indicated by M_t in Eq. (7.5).

(b) It is assumed that for any object point from which the optical wave leaves, the XP ideally creates a spherical wave converging to the image point. Thus, the pupil function is defined relative to an ideal spherical wavefront. Complex exponential terms are included in the pupil function to describe wavefront deviations from a sphere (covered in Chapter 8).

(c) The term *diffraction limited* is applied to a system with a perfect pupil function where only the boundaries of the pupil are involved with the diffractive effects.

7.2.2 Coherent transfer function examples

For the first example, consider a square pupil function given by

$$P(x, y) = \text{rect}\left(\frac{x}{2w_{XP}}\right) \text{rect}\left(\frac{y}{2w_{XP}}\right). \qquad (7.8)$$

From Eq. (7.7) the coherent transfer function is given by

$$H(f_U, f_V) = \text{rect}\left(\frac{-\lambda z_{XP} f_U}{2w_{XP}}\right)\text{rect}\left(\frac{-\lambda z_{XP} f_V}{2w_{XP}}\right). \quad (7.9)$$

Since the rectangle function is symmetric, the negative signs can be ignored. The *coherent cutoff frequency* along the u or v direction is defined as

$$f_0 = \frac{w_{XP}}{\lambda z_{XP}}. \quad (7.10)$$

Spatial frequencies with absolute values greater than f_0 will not be preserved in the image plane field.

A second example is the circular pupil function given by

$$P(x, y) = \text{circ}\left(\frac{\sqrt{x^2 + y^2}}{w_{XP}}\right), \quad (7.11)$$

where $w_{XP} = D_{XP}/2$. The coherent transfer function is

$$H(f_U, f_V) = \text{circ}\left(\frac{\sqrt{(-\lambda z_{XP} f_U)^2 + (-\lambda z_{XP} f_V)^2}}{w_{XP}}\right), \quad (7.12)$$

or

$$H(f_U, f_V) = \text{circ}\left(\frac{\sqrt{f_U^2 + f_V^2}}{f_0}\right), \quad (7.13)$$

where f_0 is again the coherent cutoff frequency as defined in Eq. (7.10). Unlike the square aperture, the cutoff frequency in this case is the same radially in all directions in the frequency plane.

To observe or record a coherent image, the irradiance given by $I_i = |U_i|^2$ is actually measured. As a result of the squaring operation, the irradiance image can theoretically gain up to *twice* the frequency content of the field—think about the fact that $\cos^2(2\pi bx) = \frac{1}{2}[1 + \cos(2\pi 2bx)]$. So, when an irradiance image is formed, the following cutoff should be considered:

$$2f_0 = \frac{2w_{XP}}{\lambda z_{XP}}. \tag{7.14}$$

Given that $2w_{XP} = D_{XP}$, and using Eq. (7.3),

$$2f_0 = \frac{1}{\lambda(f/\#)}. \tag{7.15}$$

7.2.3 Diffraction-limited coherent imaging simulation

One approach for simulating coherent imaging on the computer is based on Eq. (7.4) and implemented as

$$U_i(u,v) = \mathfrak{I}^{-1}\{H(f_U, f_V)\mathfrak{I}\{U_g(u,v)\}\}. \tag{7.16}$$

A simulation begins with a sampled "ideal" image, usually an image file that is opened in the script. The sampled ideal image is assumed to have a physical sample interval Δu and side length L. The highest spatial frequency available in the ideal image is the Nyquist frequency $f_N = 1/(2\Delta u)$, so a diffraction-limited simulation requires

$$2f_0 \leq f_N. \tag{7.17}$$

This condition comes about because you can't model spatial frequencies in the simulation that are not present in the ideal image. Applying Eq. (7.15), substituting for f_N and rearranging gives a criterion for the sample interval

$$\Delta u \leq \frac{\lambda(f/\#)}{2}. \tag{7.18}$$

Given $L = M\Delta u$, where M is the number of samples, then

$$L \leq M\frac{\lambda(f/\#)}{2}. \tag{7.19}$$

The constraint in Eq. (7.17) can be rather restrictive. Let's look at an example. Consider a thin lens with a diameter of 12.5 mm, a focal length of 125 mm, and in which the wavelength of interest is 0.5 µm. Assume a 250 × 250 sample array for an imaging simulation. First, calculate the *f*-number:

$$f/\# = \frac{125}{12.5} = 10. \tag{7.20}$$

Inserting the values in Eq. (7.18) yields

$$\Delta u \le \frac{0.5\,\mu\text{m} \cdot 10}{2} = 2.5\,\mu\text{m}, \tag{7.21}$$

and the side length constraint is, therefore,

$$L \le 250 \times 2.5\,\mu\text{m} = 0.625\,\text{mm}. \tag{7.22}$$

Thus, the image plane size is limited to 0.625 × 0.625 mm—a relatively small area. Working with a larger array increases the image size. For example, with a 2048 × 2048 sample array, the maximum image plane size would be 5.12 × 5.12 mm—still a relatively small area. This illustrates that the Fourier optics-based simulation described here examines a small part of the image plane for near-diffraction-limited performance. However, a very large array is needed to model a modest field of view, which might correspond to an image size of, say, 10 or 20 mm in this case.

Let's work up some code. The first thing we need is the ideal image. I use a 250 × 250 pixel .png image file that depicts a USAF 1951 resolution test chart. Actual test charts are printed on a glass substrate, and the USAF 1951 is still used today for testing lenses and optical systems. This file can be downloaded at http://www.ece.nmsu.edu/~davvoelz/cfo/. Similar images of the chart can also be found on the internet, or you can use another image file of your choosing, although preferably something with a variety of feature sizes.

In MATLAB start a new file named "coh_image" and enter the following:

```
1  % coh_image Coherent Imaging Example
2
3  A=imread('USAF1951B250','png');  %read image file
4  [M,N]=size(A);              %get image sample size
5  A=flipud(A);                %reverse row order
6  Ig=single(A);               %integer to floating
7  Ig=Ig/max(max(Ig));         %normalize ideal image
8
9  ug=sqrt(Ig);                %ideal image field
10 L=0.3e-3;                   %image plane side length (m)
11 du=L/M;                     %sample interval (m)
12 u=-L/2:du:L/2-du; v=u;
13
14 figure(1)                   %check ideal image
15 imagesc(u,v,Ig);
16 colormap('gray');  xlabel('u (m)'); ylabel('v (m)');
17 axis square
18 axis xy
```

In this code, `imread` loads the image file USAF1951B250.png into the temporary array A. As coded here, the image file needs to be resident in the current directory. This code assumes the image is a grayscale image ($M \times M \times 1$) as opposed to an RGB image ($M \times M \times 3$). If you use your own image with a different format, the format may need to be converted or the script set up to use it appropriately. Also, note that later code in `coh_image` assumes the ideal image is square in format with an even number of samples along a side. Changes such as defining separate u and v coordinate vectors are needed if the image is not square.

Since an image is conventionally stored with the top row first, the `flipud` function is used to reverse the order of the rows of A so the bottom of the image corresponds to the negative v coordinates. The `single` command converts the .png integer pixel values to floating point. The `max` command, applied twice to find the maximum value of a 2D array, is used to set the peak image value to unity for reference.

Since the image file represents an irradiance image, take the square root to get the magnitude of the field. This actually has no effect on this particular test chart image since after normalization it only contains zeros and ones. Beyond that, a phase component can be included to simulate a complex coherent field, but that comes in the next section. For now, zero phase is assumed across the ideal image.

The side length is $L = 0.3$ mm, which satisfies Eq. (7.22), and with $M = 250$ the sample interval is $\Delta u = 1.2 \times 10^{-6}$ m. The ideal test chart image is shown in Fig. 7.5.

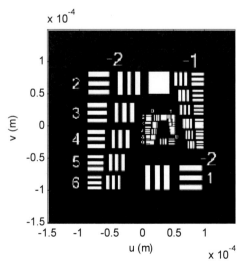

Figure 7.5 USAF test chart ideal image.

Continue with `coh_image` and add the following code to define the imaging system parameters and generate the coherent transfer function:

```
19  lambda=0.5*10^-6;        %wavelength
20  wxp=6.25e-3;             %exit pupil radius
21  zxp=125e-3;              %exit pupil distance
22  f0=wxp/(lambda*zxp);     %cutoff frequency
23
24  fu=-1/(2*du):1/L:1/(2*du)-(1/L); %freq coords
25  fv=fu;
26  [Fu,Fv]=meshgrid(fu,fv);
27  H=circ(sqrt(Fu.^2+Fv.^2)/f0);
28
29  figure(2)                %check H
30  surf(fu,fv,H.*.99)
31  camlight left; lighting phong
32  colormap('gray')
33  shading interp
34  ylabel('fu (cyc/m)'); xlabel('fv (cyc/m)');
```

Visible wavelength illumination is assumed, as is a pupil radius of 6.25 mm with an XP distance of 125 mm. For the $f/10$ lens the following are computed:

$$f_N = 1/(2 \cdot 1.2 \times 10^{-6}) = 4.17 \times 10^5 \text{ cycles/m},$$

$$2f_0 = 1/(0.5 \times 10^{-6} \cdot 10) = 2 \times 10^5 \text{ cycles/m},$$

$$f_0 = 1 \times 10^5 \text{ cycles/m}.$$

Thus, Eq. (7.17) is satisfied. The coherent transfer function is displayed in Fig. 7.6. If Eq. (7.17) were violated, the pupil function in Fig. 7.6 would reach beyond halfway to the array boundaries. The combination of surface plotting and lighting

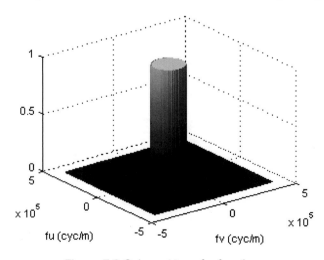

Figure 7.6 Coherent transfer function.

commands in the script help improve the display of H. In MATLAB Version 7.1 the surf (or mesh) commands have trouble displaying circle functions of unit height. Changing the height slightly (* 0.99) allows the plot to display.

Add the following to generate the simulated image:

```
35  H=fftshift(H);
36  Gg=fft2(fftshift(ug));
37  Gi=Gg.*H;
38  ui=ifftshift(ifft2(Gi));
39  Ii=(abs(ui)).^2;
40
41  figure(3)              %image result
42  imagesc(u,v,nthroot(Ii,2));
43  colormap('gray'); xlabel('u (m)'); ylabel('v (m)');
44  axis square
45  axis xy
46
47  figure(4)              %horizontal image slice
48  vvalue=-0.8e-4;        %select row (y value)
49  vindex=round(vvalue/du+(M/2+1)); %convert row index
50  plot(u,Ii(vindex,:),u,Ig(vindex,:),':');
51  xlabel('u (m)'); ylabel('Irradiance');
```

In this piece of code Eq. (7.16) is implemented. The resulting image is displayed in Fig. 7.7, where the square root of the irradiance nthroot(Ii,2) is applied to boost the contrast. The features are blurred and some of the three-bar groups are unresolved. Constructive and destructive interference produce "ringing" features.

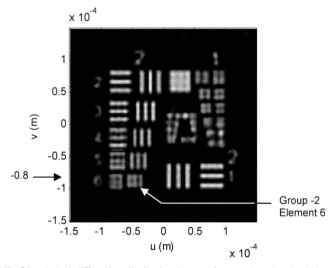

Figure 7.7 Simulated diffraction-limited coherent image (contrast enhanced).

The last portion of the code displays a profile of a row in the image. A physical *v*-axis coordinate value (vvalue) is selected, and the nearest array index value is found. The round function rounds a floating point value to the nearest integer. The plot compares the ideal and simulated image profiles.

In Fig. 7.7 the three bars of Group -2, Element 6, appear to be the smallest group that is "resolved." Selecting the *v*-coordinate of -0.8×10^{-4} m in the profile code gives the display in Fig. 7.8. The large bars of Group -2, Element 1 are clearly resolved but with some obvious ringing effects. Element 6 vertical bars are resolved but with much less contrast.

Some resolution "sleuthing" can be done as a sanity check. Expand the profile plot and observe that the Element 6 bars have a period of roughly 1×10^{-5} m. Inverting the period gives a corresponding spatial frequency of $\sim 1 \times 10^{5}$ cycles/m. This is the same as the coherent cutoff $f_0 = 1 \times 10^5$ cycles/m. Even though the square magnitude of the field is taken for the irradiance, the apparent resolution in this kind of simulated coherent irradiance image, typically, appears close to the coherent cutoff.

Figure 7.9 shows a sequence of image spectra with the coherent transfer function and resulting irradiance images for different pupil sizes/*f*-numbers. The log of the spectra magnitude is displayed to bring out low contrast details.

7.2.4 Rough object

In the previous section the test chart was assumed to be a "binary" object where the source field amplitude is either zero or unity. Furthermore, the phase of the source field across the chart was constant (equal to zero). However, when a

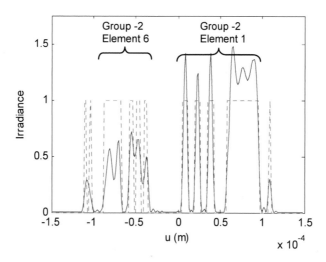

Figure 7.8 Row profile of simulated coherent image and ideal image at $v = -0.8 \times 10^{-4}$ m.

Imaging and Diffraction-Limited Imaging Simulation

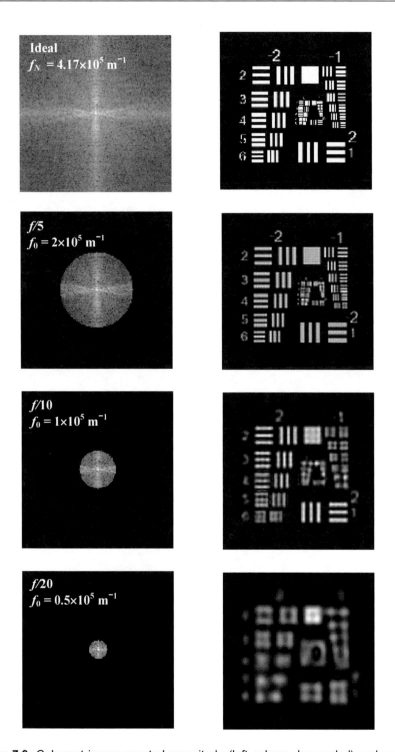

Figure 7.9 Coherent image spectral magnitude (left column, log scaled) and associated irradiance images (right column). Image scale is the same as Fig. 7.7.

physical object is illuminated with coherent light, its micro-surface properties can have a considerable effect on the diffracted field. Most surfaces are "rough" relative to the size of the optical wavelength. In other words, the random dips and bumps of a surface, unless it is machined like a mirror, are large enough that an incident plane wave reflecting from one dip will have a significantly different phase from the portion of the wave that reflects off a bump.

A simple way to model this effect is to apply a random complex exponential phase term to the object (ideal image) field. Make the following change to `ug` in `coh_image`:

```
ug=sqrt(Ig).*exp(j*2*pi*rand(M));   %ideal image
```

The `rand` function produces an M×M array of random values of uniform distribution over the interval [0,1]. Multiplying by 2π scales the values to range over all possible phase values [0,2π]. Therefore, every sample point has a phase that is independent and uncorrelated from every other point. This phase function represents a random surface that is "extremely rough" relative to the optical wavelength.

The code now produces the irradiance image of Fig. 7.10. The coherent interference of the wavelets coming off the rough surface produce what is known as "speckle," a random jumble of spots, dots, and squiggles. Coherent speckle is a well-known phenomenon.[5] You can experience speckle effects for yourself by shining a laser pointer at a wall and moving the spot slightly, or by shifting your head. The sparkling is a speckle effect. Figure 7.10 illustrates that speckle can seriously impede the ability to interpret, or gain, information from an image.

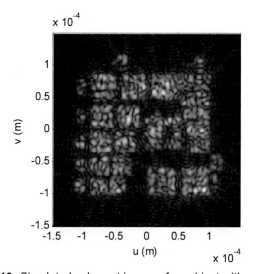

Figure 7.10 Simulated coherent image of an object with a rough surface.

7.3 Incoherent Imaging

7.3.1 Incoherent imaging theory

Imagine that an object is represented by a collection of randomly radiating point sources. The object may actually be illuminated by a source, like the sun, but ultimately the field exiting the object surface involves a spectrum of wavelengths and randomly changing phase in time. Interference features produced by this type of source are not stationary but change rapidly, so they are "averaged out" by a sensor responding to the time-averaged squared magnitude of the field. Perfectly *incoherent* light refers to the situation where the complex field phasors from the radiating point sources are stochastically independent; where there is no correlation between the field phasors at different points or times. To visualize this idea, imagine the speckles in the image of Fig. 7.10 changing randomly and rapidly in time. With enough averaging the image texture will tend to become smooth.

Assuming incoherent illumination of an object, the linear, space invariant model for imaging becomes[4]

$$I_i(u,v) = |h(u,v)|^2 \otimes I_g(u,v), \qquad (7.23)$$

where h is the same coherent impulse response indicated in Eq. (7.4) and I_g is the ideal geometric irradiance image. In contrast to coherent imaging, which is linear with the *field*, incoherent imaging is linear with *irradiance*. The impulse response $|h(u,v)|^2$ is commonly known as the *point spread function* (PSF). The discussion here is restricted to light with only a small wavelength spread around a center wavelength of λ.

The corresponding spectra of the functions in Eq. (7.23) are related by

$$G_i(f_U, f_V) = \mathcal{H}(f_U, f_V) G_g(f_U, f_V), \qquad (7.24)$$

where $\mathcal{H}$ is known as the *optical transfer function* (OTF). By convention, the OTF is normalized as follows:

$$\mathcal{H}(f_U, f_V) = \frac{\mathfrak{I}\{|h(u,v)|^2\}}{\int\int_{-\infty}^{\infty} |h(u,v)|^2 \, du\, dv}. \qquad (7.25)$$

Considering the Fourier autocorrelation theorem for the numerator, the OTF is a normalized autocorrelation of the coherent transfer function $H(f_U, f_V)$. The normalization amounts to scaling the OTF to have a value of 1 at the DC frequency, $(f_U, f_V) = (0,0)$. This means the OTF is not assumed to affect the total optical power associated with the ideal geometrical image.

In shorthand notation, where ★ indicates a correlation, the OTF can be written as

$$\mathcal{H}(f_U, f_V) = H(f_U, f_V) \star H(f_U, f_V)|_{norm}. \tag{7.26}$$

7.3.2 Optical transfer function examples

Starting with the coherent transfer function of Eq. (7.9) for the square pupil, the Fourier autocorrelation theorem can be applied to find the OTF. First, find the Fourier transform then take the squared modulus:

$$|\Im\{H(f_U, f_V)\}|^2 = \left(\frac{4w_{XP}^2}{\lambda^2 z_{XP}^2}\right)^2 \text{sinc}^2\left(\frac{2w_{XP}}{\lambda z_{XP}} u\right) \text{sinc}^2\left(\frac{2w_{XP}}{\lambda z_{XP}} v\right). \tag{7.27}$$

The inverse transform of Eq. (7.27) gives the autocorrelation function:

$$H(f_U, f_V) \star H(f_U, f_V) = \frac{4w_{XP}^2}{\lambda^2 z_{XP}^2} \Lambda\left(\frac{\lambda z_{XP}}{2w_{XP}} f_U\right) \Lambda\left(\frac{\lambda z_{XP}}{2w_{XP}} f_V\right). \tag{7.28}$$

Removing the front multiplying terms gives a normalized function with unit value at zero frequency, so the OTF is

$$\mathcal{H}(f_U, f_V) = \Lambda\left(\frac{\lambda z_{XP}}{2w_{XP}} f_U\right) \Lambda\left(\frac{\lambda z_{XP}}{2w_{XP}} f_V\right), \tag{7.29}$$

where the *incoherent cutoff frequency* for this OTF is

$$2f_0 = \frac{2w_{XP}}{\lambda z_{XP}}, \tag{7.30}$$

which is twice the coherent cutoff frequency for this aperture. The incoherent cutoff is the same as the cutoff suggested in Eq. (7.14).

What about the circular pupil function for the coherent example in Section 7.2.2? For its OTF, the normalized autocorrelation of the coherent transfer function in Eq. (7.13) is needed. This is not easy to find directly, but a graphical correlation approach is applied in this case with the result[4]

$$\mathcal{H}(f_U, f_V) = \begin{cases} \dfrac{2}{\pi}\left[\arccos\left(\dfrac{\rho}{2\rho_0}\right) - \left(\dfrac{\rho}{2\rho_0}\right)\sqrt{1-\left(\dfrac{\rho}{2\rho_0}\right)^2}\right], & \rho \leq 2\rho_0 \\ 0, & \text{otherwise} \end{cases} \quad (7.31)$$

where $\rho = \sqrt{f_U^2 + f_V^2}$ and the incoherent cutoff is given by

$$2\rho_0 = \frac{2w_{XP}}{\lambda z_{XP}} = \frac{1}{\lambda f/\#}. \quad (7.32)$$

7.3.3 Diffraction-limited incoherent imaging simulation

An incoherent image simulation based on Eq. (7.23) can be implemented as

$$I_i(u,v) = \mathfrak{I}^{-1}\left\{\mathcal{H}(f_U, f_V)\mathfrak{I}\{I_g(u,v)\}\right\}. \quad (7.33)$$

The incoherent simulation follows the same form as the coherent case since the OTF is developed from the coherent transfer function. The criterion of Eq. (7.17) still applies. The following script, "incoh_image," uses the same parameters as the coherent example in Section 7.2.3:

```
1   %  incoh_image Incoherent Imaging Example
2
3   A=imread('USAF1951B250','png');  %read image file
4   [M,N]=size(A);           %get image sample size
5   A=flipud(A);             %reverse row order
6   Ig=single(A);            %integer to floating
7   Ig=Ig/max(max(Ig));      %normalize ideal image
8
9   L=0.3e-3;                %image plane side length (m)
10  du=L/M;                  %sample interval (m)
11  u=-L/2:du:L/2-du; v=u;
12
13  lambda=0.5*10^-6;        %wavelength
14  wxp=6.25e-3;             %exit pupil radius
15  zxp=125e-3;              %exit pupil distance
16  f0=wxp/(lambda*zxp);     %coherent cutoff
17
18  fu=-1/(2*du):1/L:1/(2*du)-(1/L); %freq coords
19  fv=fu;
20  [Fu,Fv]=meshgrid(fu,fv);
21  H=circ(sqrt(Fu.^2+Fv.^2)/f0);
22  OTF=ifft2(abs(fft2(fftshift(H))).^2);
23  OTF=OTF/OTF(1,1);
24
```

```
25  figure(2)                %check OTF
26  surf(fu,fv,fftshift(abs(OTF)))
27  camlight left; lighting phong
28  colormap('gray')
29  shading interp
30  ylabel('fu (cyc/m)'); xlabel('fv (cyc/m)');
31
32  Gg=fft2(fftshift(Ig)); %convolution
33  Gi=Gg.*OTF;
34  Ii=ifftshift(ifft2(Gi));
35  %remove residual imag parts, values <  0
36  Ii=real(Ii); mask=Ii>=0; Ii=mask.*Ii;
37
38  figure(3)                %image result
39  imagesc(u,v,nthroot(Ii,2));
40  colormap('gray'); xlabel('u (m)'); ylabel('v (m)');
41  axis square;
42  axis xy;
43
44  figure(4)                %horizontal image slice
45  vvalue=0.2e-4;           %select row (y value)
46  vindex=round(vvalue/du+(M/2+1)); %convert row index
47  plot(u,Ii(vindex,:),u,Ig(vindex,:),':');
48  xlabel('u (m)'); ylabel('Irradiance');
```

Some comments on this code:

(a) Line 7: I_g is required rather than U_g.
(b) Line 22: The OTF is calculated by applying the autocorrelation theorem.
(c) Line 23: The OTF is normalized by the zero-frequency value of the autocorrelation result. Note that the OTF is left in the shifted arrangement.
(d) Line 36: The computed I_i should be positive and real valued. But, numerical precision can produce small imaginary and negative values in the results. This is not a big problem except that the `imagesc` and `plot` routines don't like them—so they are set to zero in the script.

The diffraction-limited OTF is shown in Fig. 7.11 and the resulting image is shown in Fig. 7.12. Compared with the coherent image of Fig. 7.7, the resolution is better (see the bars of Group −1, Element 3) and there are no ringing features.

Figure 7.13 shows an x-axis profile through the image at the row corresponding to $v = 0.2 \times 10^{-4}$ m. It appears the bars of Group −1, Element 3, are just resolved. An examination of these bars yields a spatial frequency of ~ 1.7 $\times 10^5$ cycles/m, which is slightly smaller than—but on the order of—the incoherent cutoff of 2×10^5 cycles/m. Figure 7.14 shows a set of incoherent images for the same series of *f*-numbers used in the coherent image display of Fig. 7.9. Incoherent light certainly improves the image quality in this case.

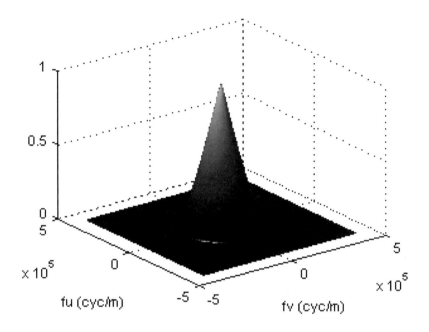

Figure 7.11 Incoherent transfer function.

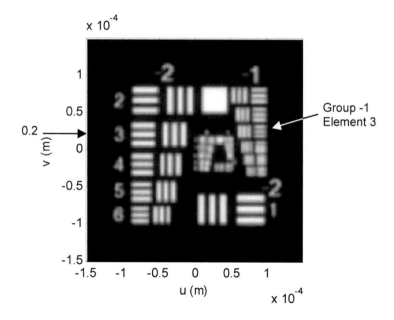

Figure 7.12 Simulated diffraction-limited incoherent image (contrast enhanced).

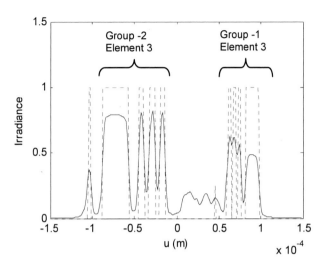

Figure 7.13 Row profile of simulated incoherent image and ideal image at $v = 0.2 \times 10^{-4}$ m.

7.4 Exercises

7.1 Consider a thin lens of radius 10 mm and focal length $f = 50$ mm. An object is positioned 200 mm from the lens.

(a) What is the image distance (back principal plane to image location)? What is the transverse magnification?

(b) What are D_{XP} and z_{XP}? What is the $f/\#$?

(c) Suppose the object distance is increased such that $z_1 \gg f$, what is the approximate image distance?

7.2 Suppose $D_{XP} = 1/2$ in. and $z_{XP} = 4$ in. for a wavelength of 0.587 μm.

(a) What is the $f/\#$? What are the coherent and incoherent cutoff frequencies?

(b) What is the sample interval requirement for an ideal image in a diffraction-limited simulation?

(c) If the ideal image size corresponds to a side length of 1 mm, what is the requirement for the number of samples across the ideal image array in a simulation?

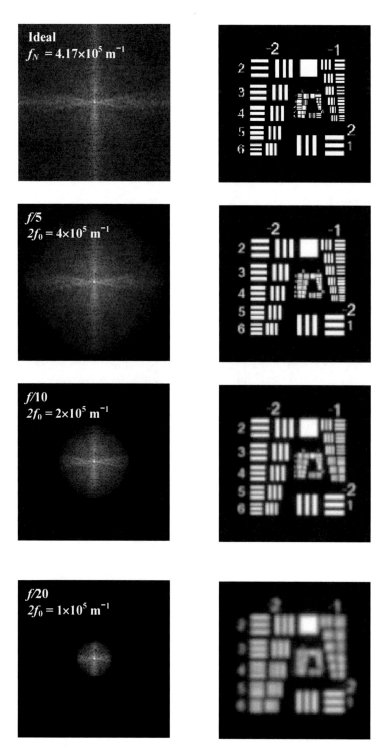

Figure 7.14 Incoherent image spectral magnitude (left column, fifth root scaled) and associated irradiance images (right column). Image scale is the same as Fig. 7.11.

7.3 Assume $z_{XP} = 50$ mm, $\lambda = 0.5$ μm, and a rectangular XP with half widths of $w_x = 1$ mm and $w_y = 0.5$ mm. Use $L = 1$ mm and $M = 250$.

(a) What are the coherent and incoherent cutoff frequencies (x and y directions)?

(b) Develop a diffraction-limited coherent image simulation with the test chart as the image. Present a surface plot of the coherent transfer function and the simulated coherent image.

(c) Develop a diffraction-limited incoherent image simulation with the test chart as the image. Present a surface plot of the OTF and the simulated incoherent image.

(d) Is a spatial resolution difference in the x and y directions apparent?

7.4 Assume $z_{XP} = 50$ mm, $\lambda = 0.5$ μm, and an annular XP with an outer radius of $w_o = 1$ mm and an inner radius of $w_i = 0.5$ mm. Use $L = 1$ mm and $M = 250$.

(a) What are the coherent and incoherent cutoff frequencies?

(b) Develop a diffraction-limited coherent image simulation with the test chart as the image. Present a surface plot of the coherent transfer function and the simulated coherent image.

(c) Develop a diffraction-limited incoherent image simulation with the test chart as the image. Present a surface plot of the OTF and the simulated incoherent image.

(d) What is the effect of "losing" the DC and low spatial frequency components with the coherent image transfer function?

7.5 Concerning the rough object coherent image simulation in Section 7.2.4:

(a) Suppose the object is "not so rough" compared to the optical wavelength. Reduce the 2π multiplier for the random number function. What is the effect on the irradiance image? Explain the result.

(b) What is the effect on the irradiance image of changing the aperture size? Explain the result.

7.6 Diffraction-limited PSF (Airy pattern): Consider the coherent image transfer function $H(f_u, f_v)$ for a circular aperture given in Eq. (7.13).

(a) Derive an expression for the impulse response $h(u,v)$.

(b) Find an expression for the incoherent PSF. Show that this result is of the same form as the Airy pattern derived in Eq. (4.35).

(c) Show that the full width of the central lobe of the Airy pattern PSF is equal to $2.44\lambda(f/\#)$.

7.7 Rayleigh resolution criterion: Spatial frequency cutoff values imply something about the diffraction-limited resolution for an imaging system. However, there are a number of specific metrics for characterizing resolution, for example, the PSF and modulation transfer function (MTF) to be discussed in Chapter 8. Another metric, the Rayleigh criterion, postulates that two incoherent point sources are "just resolved" by a diffraction-limited system with a circular pupil when the point images are separated by half the center lobe width of the Airy pattern. The full width of the Airy pattern is $2.44\lambda f/\#$ (see Exercise 7.6). Taking half this value gives the Rayleigh criterion distance as

$$\delta = 1.22\lambda(f/\#).$$

Test this expression empirically with a simulation based on the example incoherent system of Section 7.3.3. Choose w_{XP} such that $f/\# = 20$.

(a) Calculate the value of δ.

(b) For the image plane sample interval Δu in the example, how many samples S are required to span the distance δ?

(c) Create an ideal image frame consisting of two point sources separated by δ. It is easier in this case to work with the array indices rather than the physical units. Try some lines like

```
Ig=zeros(M);
Ig(M/2+1,M/2+1-S/2)=1;
Ig(M/2+1,M/2+1+S/2)=1;
```

(d) Run the image simulation. Be sure to choose w_{XP} such that $f/\# = 20$. Are the two points resolved? Plot a profile through the two-point irradiance pattern.

7.8 Object space: In some situations it is convenient to discuss the spatial frequency cutoff or image resolution in terms of the *object coordinates*. For example, the resolution of Earth-imaging satellites is discussed in terms of resolution at the Earth's surface, not at the image plane. To convert the cutoff and resolution expressions from an image plane reference to an object plane reference, simply substitute the object space *f*-number ($f/\#_{OBJ}$) for the image space *f*-number ($f/\#$). The object space *f*-number is

$$f/\#_{OBJ} = \frac{z_{EP}}{D_{EP}}.$$

Consider a telescope on a satellite orbiting the Earth at an altitude of 700 km. The telescope is directed toward the surface and has a primary mirror diameter of 1 m. The mirror is the entrance pupil. Assume $\lambda = 0.5$ μm.

(a) What is $f/\#_{OBJ}$?

(b) Determine the incoherent cutoff frequency. Find a value for the Rayleigh resolution distance δ (see Exercise 7.7).

(c) What is the maximum sample interval that can be used in a simulation of the object plane? What is the corresponding object plane side length for a 1024 × 1024 array?

7.9 Phase Contrast Imaging: Objects such as living cells, microorganisms, and lithographic patterns are essentially transparent, although their features can have differing indices of refraction. A conventional image of this type of object will not reveal the features. However, a phase contrast imaging method, often used in microscopy, converts the small phase changes (optical path differences) produced by the object into irradiance changes that can be detected. More details on this imaging approach can be found in Ersoy or Hecht, for example.[6,7]

A simple arrangement to demonstrate phase contrast imaging is shown in Fig. 7.15. An optical Fourier transform [see Eq. (6.19) and related discussion] is taken of the phase object with lens 1 (focal length f), and the DC part of the transformed field is altered with a phase "dot" plate. Lens 2 performs a second transform to produce the image.

The general idea is to phase shift the mean part of the object field by a fraction of a wavelength, usually $\pi/2$ or $3\pi/2$, with the phase plate. The mean field "interferes" with the phase-modulated field structure at the image plane to translate phase variations into irradiance variations. Write some code to simulate this setup following these steps:

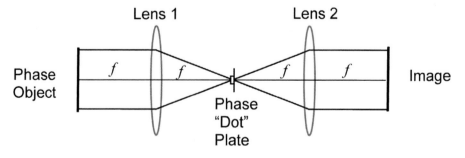

Figure 7.15 Phase contrast image arrangement.

(a) Read in the test chart `USAF1951B250.png` as in `coh_image` but convert it to a unit magnitude phase object $u_g(x,y) = \exp[j\theta(x,y)]$ with the following:

```
theta=sqrt(Ig)*pi/100;
ug=exp(j*theta);
```

(b) Directly generate the irradiance pattern of this object (simply compute the squared magnitude) and examine the result using `imagesc`. Also examine a profile. These results should show a constant value of 1. However, recall that `imagesc` stretches (or contracts) the full range of image values to be displayed in 256 gray levels. Thus, some indications of the object features may be present due to numerical precision error, but these variations are extremely small in absolute value.

(c) Assume the object is back-illuminated by a unit amplitude wave. Simulate the effect of lens 1 [i.e., compute the fast Fourier transform (FFT) of the phase object]. Coordinate and amplitude scaling can be ignored for this exercise. Multiply the DC point by $\exp(j\beta)$, where $\beta = \pi/2$. Compute the inverse FFT of the result to simulate the effect of lens 2. Lens 2 actually performs a forward transform, which produces an image that is rotated 180 deg relative to the object. However, the inverse FFT is used in the simulation to keep the image orientation the same as the object. Examine the image irradiance pattern and a profile through the pattern. The results should be approximately given by $I_i(x,y) \approx 1+2\theta(x,y)$. When $\beta = \pi/2$, this is referred to as the positive phase contrast mode.

(d) Repeat (c) but use $\beta = 3\pi/2$, which is referred to as the negative phase contrast mode. Demonstrate with profile plots that the simulation result is approximately given by $I_i(x,y) \approx 1-2\theta(x,y)$.

(e) Try other values of β. What is the effect of changing β?

7.10 Holography: Holography refers to an irradiance recording, traditionally on a piece of film, that can be "played back" to recreate the optical field radiating from an object of interest. By recreating the field, the recorded object as seen by an observer will exhibit some of the same characteristics as viewing a real object, such as 3D depth. Here, a version of a *Fourier hologram* is considered (Fig. 7.16).

For the recording, an object is illuminated with coherent light (usually a plane wave). To the side of the object a small spherical mirror intercepts some of the illumination beam and sends a spherical wave back toward the lens. The mirror, in effect, emulates a reference point source. The Fraunhofer pattern of the illuminated object/point source is created at the

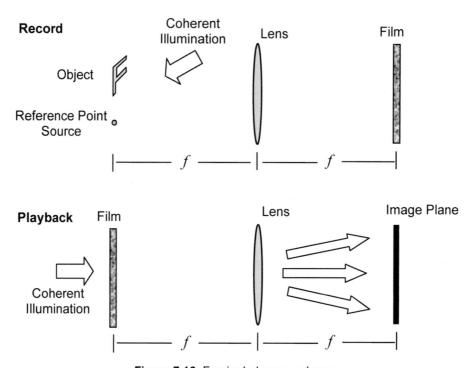

Figure 7.16 Fourier hologram scheme.

film plane by the lens. The film records the irradiance pattern and is processed so its transmittance function replicates the irradiance. For the playback, the film is illuminated by a coherent plane wave and a lens is used to form the Fraunhofer pattern of the transmitted wavefront.

Follow these steps to simulate the Fourier hologram process:

(a) Use $M = 500$, $L = 10$ mm to define the x- and y-coordinate vectors for the object array. Use shifted `rect` functions and the `udelta` function to create the object and point source in the array as diagramed in Fig. 7.17. Display an image of the object array to check your layout.

(b) Multiply the point source by 1000. A bright reference aids in creating the hologram.

(c) To simulate the effect of the recording lens, simply compute the FFT of the object array. Coordinate and amplitude scaling can be ignored for this exercise. Compute the squared magnitude of the result to emulate the irradiance recording by the film.

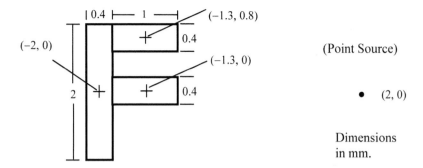

Figure 7.17 Holography example object and reference point source. Centering coordinates are indicated with parentheses.

(d) Assume the film is illuminated by a unit amplitude plane wave for playback (this requires no action on your part). To get the image plane result, compute the inverse FFT of the result from step (c). You may also need to consider `fftshift`. The playback lens actually performs a forward transform, which produces a 180-deg rotated image. However, the inverse FFT is used here to keep the image orientation the same as the object.

(e) The result of step (d) should have a bright point on what is effectively the optical axis. This point corresponds to all of the energy from the playback beam that is still planar after passing through the film. It causes display difficulties, so set it to zero.

(f) Display the result of step (e) in an image.

The Fourier hologram produces an image of the object, a "conjugate" image of the object, and an autocorrelation-related result in the middle. A few questions to consider: Do you suppose you could see only the image of the object if the playback lens were removed and you positioned your eye appropriately? Do you recognize this Fourier hologram process as simply an implementation of the Fourier autocorrelation theorem?

7.5 References

1. E. Dereniak and T. D. Dereniak, *Geometrical and Trigonometric Optics*, Cambridge University Press, Cambridge, U.K. (2008).
2. W. J. Smith, *Modern Optical Engineering*, 4th Ed., McGraw-Hill Professional, New York (2007).
3. J. M. Geary, *Introduction to Lens Design with Practical ZEMAX® Examples*, Willmann-Bell, Richmond, VA (2002).
4. J. W. Goodman, *Introduction to Fourier Optics*, 3rd Ed., Roberts & Company, Greenwood Village, CO (2005).

5. J. W. Goodman, *Speckle Phenomena in Optics*, Roberts & Company, Greenwood Village, CO (2007).
6. O. K. Ersoy, *Diffraction, Fourier Optics, and Imaging*, Wiley-Interscience, New York (2006).
7. E. Hecht, *Optics*, 4th Ed., Addison-Wesley, Reading, MA (2002).

Chapter 8
Wavefront Aberrations

The pupil function for the diffraction-limited imaging system model of Chapter 7 is defined relative to an ideal converging spherical wavefront. A system with *aberrations* has a wavefront phase surface that deviates from the ideal spherical wave. Aberrations are found in most practical imaging systems, and their effect reduces image quality. If aberrations are significant, then a ray (geometrical) optics approach typically is used for studying image effects. However, if the system is "relatively close" to diffraction limited, then wave optics can be the tool of choice. In this chapter the diffraction-limited imaging theory is extended to include aberrations. The incoherent point spread function (PSF) and modulation transfer function (MTF) are discussed and demonstrated as image quality measures. Aberrated systems tend to cause *space-variant* imaging, where the impulse response is not the same for each image point. An example of a space-variant image simulation is presented.

8.1 Wavefront Optical Path Difference

Figure 8.1 shows the exit pupil (XP) with an ideal spherical (sp) wavefront and aberrated (ab) wavefront in profile. The wavefront error is described by $W(x,y)$, an optical path difference (OPD) function that represents the difference between the spherical and aberrated wavefront surfaces. x and y are coordinates in the pupil plane.

Aberrated wavefronts arise from various sources. Obvious examples are imperfections in the imaging optics. For systems that peer through the atmosphere, the wavefront disturbances caused by the turbulence can be characterized in terms of aberrated wavefronts.

However, even when optical components are made exactly to specification, aberrations will be present. For example, chromatic aberration, where different wavelengths focus at different positions, is caused by the wavelength dependence of the index of refraction of glass. Furthermore, if a system images an extended object scene, light from off-axis points must transit the system at an angle relative to the optical axis, and this generates a departure from a spherical wave.

Wavefront OPD is commonly described by a polynomial series. The Seidel series is used by optical designers because the terms have straightforward mathematical relationships to factors such as lens type and position in the image

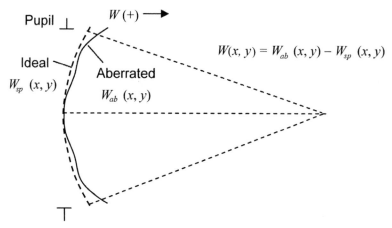

Figure 8.1 Spherical (sp) and aberrated (ab) wavefronts.

plane. Another series, Zernike polynomials, is used in optical testing and applications where the aberrations do not have a simple dependency on the system parameters. This series possesses useful properties such as orthogonal terms. Both formulations assume a circular pupil.

In this chapter wavefront OPD is incorporated in the image system pupil function. Monochromatic aberrations are considered, where color is not involved. The topic is initially approached by considering the primary Seidel aberrations.

8.2 Seidel Polynomials

8.2.1 Definition and primary aberrations

Seidel polynomials are often used to describe monochromatic aberrations for rotationally symmetric optical systems, such as most lenses and mirrors. A common form that is applied in conventional imaging systems is described by[1,2]

$$W(\hat{u}_0; \rho, \theta) = \sum_{j,m,n} W_{klm} \hat{u}_0^k \rho^l \cos^m \theta; \quad k = 2j+m, \ l = 2n+m, \quad (8.1)$$

where ρ is a *normalized radial distance in the XP* and θ is the *angle in the XP* as shown in Fig. 8.2(a). For computational reasons the angle θ is defined here relative to the x axis in a counter-clockwise direction. However, note that this angle is often defined relative to the y axis in traditional aberration treatments.[2] The normalized XP has a radius of 1 where the physical coordinates (x, y) are divided by the XP radius to get normalized coordinates $(\hat{x}, \hat{y})$. $\hat{u}_0$ is a *fractional image height*, or normalized image height, defined along the $\hat{u}$ axis in the imaging plane as indicated in Fig. 8.2(b). The fractional image height is the physical height of a given point in the image divided by the maximum image radius being considered.

Wavefront Aberrations

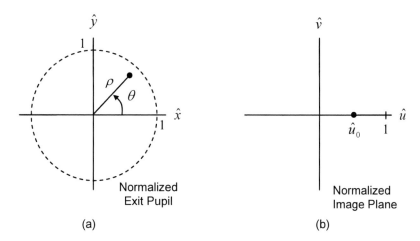

Figure 8.2 Seidel aberration coordinate definitions for the normalized (a) exit pupil and (b) image plane.

Since the Seidel polynomials assume a rotationally symmetric system, the pupil and image plane coordinate systems are simply rotated to find the wavefront OPD function for an image point that is off the $\hat{u}$ axis.

The indices j, m, n, and so forth, in Eq. (8.1) are a numbering and power scheme. W_{klm} are the wavefront aberration coefficients, and the five primary Seidel aberrations correspond to $k + l = 4$. These primary aberrations are known as *spherical aberration, coma, astigmatism, field curvature,* and *distortion*. The coefficients have units of distance (μm), although they are usually discussed relative to the optical wavelength (i.e., so many "*waves*"). Our interest is in simulating the effects of these aberrations, so we refer the reader to other resources, for example, references 1–4, for more specific discussion and insight regarding the primary aberrations.

For simulation purposes it is convenient to convert from polar to Cartesian coordinates. Referring to Fig. 8.2,

$$\rho = \sqrt{\hat{x}^2 + \hat{y}^2} \quad \text{and} \quad \rho\cos\theta = \hat{x},$$

and the primary aberrations are then written

$$\begin{aligned} W(\hat{u}_0; \hat{x}, \hat{y}) = &W_d\left(\hat{x}^2 + \hat{y}^2\right) + W_{040}\left(\hat{x}^2 + \hat{y}^2\right)^2 \\ &+ W_{131}\hat{u}_0\left(\hat{x}^2 + \hat{y}^2\right)\hat{x} + W_{222}\hat{u}_0^2\hat{x}^2 \\ &+ W_{220}\hat{u}_0^2\left(\hat{x}^2 + \hat{y}^2\right) + W_{311}\hat{u}_0^3\hat{x}. \end{aligned} \quad (8.2)$$

The first term in this series is not one of the five primary aberrations, but is a

Table 8.1 Seidel aberrations.

Aberration	Coefficient	$W(x,y)$ term
Defocus	W_d	$W_d(\hat{x}^2 + \hat{y}^2)$
Spherical	W_{040}	$W_{040}(\hat{x}^2 + \hat{y}^2)^2$
Coma	W_{131}	$W_{131}\hat{u}_0(\hat{x}^2 + \hat{y}^2)\hat{x}$
Astigmatism	W_{222}	$W_{222}\hat{u}_0^2\hat{x}^2$
Field curvature	W_{220}	$W_{220}\hat{u}_0^2(\hat{x}^2 + \hat{y}^2)$
Distortion	W_{311}	$W_{311}\hat{u}_0^3\hat{x}$

defocus term. It is the wavefront OPD that is "created" in moving the image plane along the optical axis from the paraxial focus position.

Table 8.1 names and lists the Seidel aberration terms. Spherical aberration is caused by using a spherical surface (lens or mirror surface) to converge light. It may seem this should yield a spherical wavefront but, in fact, it doesn't. The other aberrations are functions of the fractional image height. An examination of the expressions shows each coefficient is equal to the OPD at the edge of the pupil (where $\hat{x}^2 + \hat{y}^2 = 1$ and/or $\hat{x} = 1$) for the wavefront that is traveling to the maximum height image position ($\hat{u}_0 = 1$).

8.2.2 MATLAB function

The following function "seidel_5" evaluates $W(\hat{u}_0, \hat{v}_0; \hat{x}, \hat{y})$:

```
1    function[w]=seidel_5(u0,v0,X,Y,...
2            wd,w040,w131,w222,w220,w311)
3    % seidel_5
4    % Compute wavefront OPD for first 5 Seidel wavefront
5    % aberration coefficients + defocus
6    %
7    %
8    % u0,v0 - normalized image plane coordinate
9    % X,Y - normalized pupil coordinate arrays
10   %       (like from meshgrid)
11   % wd-defocus; w040-spherical; w131-coma;
12   % w222-astigmatism; w220-field curvature;
13   % w311-distortion
14
15   beta=atan2(v0,u0);     % image rotation angle
16   u0r=sqrt(u0^2+v0^2);   % image height
17
```

```
18  % rotate grid
19  Xr=X*cos(beta)+Y*sin(beta);
20  Yr=-X*sin(beta)+Y*cos(beta);
21
22  % Seidel polynomials
23  rho2=Xr.^2+Yr.^2;
24  w=wd*rho2+...
25      w040*rho2.^2+...
26      w131*u0r*rho2.*Xr+...
27      w222*u0r^2*Xr.^2+...
28      w220*u0r^2*rho2+...
29      w311*u0r^3*Xr;
30  end
```

Inputs to this function are the image plane point of interest given by the normalized coordinate pair $(\hat{u}_0, \hat{v}_0)$, the normalized pupil coordinates in the form of the X and Y arrays like those generated by meshgrid, and the aberration coefficients. The coefficients could easily be put into a vector, but are listed here for simplicity. In lines 19 and 20, a rotation transformation, where $\beta = \tan^{-1}(\hat{v}_0 / \hat{u}_0)$, is applied to align the pupil coordinate system with the image plane point. The routine calculates values for the full grid of pupil coordinates, even beyond a radius of 1. These extraneous values can be removed later.

The following piece of code can be used to create surface plots to visualize the wavefront functions for different inputs. Several examples are presented in Fig. 8.3. Note the use of the "not a number" (NaN) function and logical indexing to remove the values outside the unit circle pupil.

```
u0=1; v0=0;
wd=0; w040=1; w131=0; w222=0; w220=0; w311=0;
w=seidel_5(u0,v0,X,Y,wd,w040,w131,w222,w220,w311);
P=circ(sqrt(X.^2+Y.^2));
mask=(P==0);
w(mask)=NaN;

figure(1)
surfc(x,y,w)
camlight left; lighting phong;
colormap('gray'); shading interp;
xlabel('x'); ylabel('y');
```

Try plotting some wavefront OPD surfaces. Choose a reasonable number of samples and a side length of 2. Figure 8.3 illustrates that spherical aberration (W_{040}) and field curvature (W_{220}) are wavefront curvature-like terms that are spherically symmetric with respect to the pupil coordinates. Coma (W_{131}) and astigmatism (W_{222}) are not spherically symmetric and depend on the image point position. A few more comments about the primary Seidel wavefront OPD function follow:

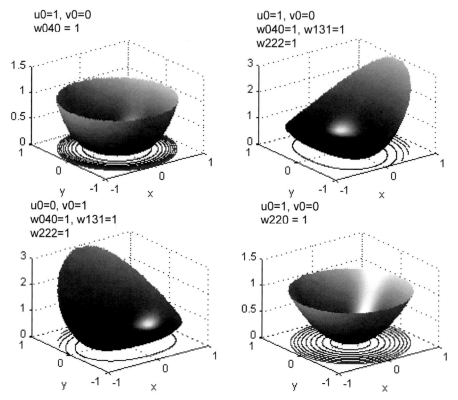

Figure 8.3 Example wavefront OPD surface and contour plots using seidel_5. These represent phase surfaces that are applied in the exit pupil.

(a) The axial image point ($\hat{u}_0 = 0, \hat{v}_0 = 0$) is only affected by spherical aberration or defocus.

(b) Systems usually have more than one aberration term because of the way aberrations arise from spherical surfaces.

(c) The primary aberrations describe most of the wavefront OPD for a conventional lens system, but higher-order terms in the Seidel polynomials can be significant in some cases.

8.3 Pupil and Transfer Functions

8.3.1 Pupil function

Applying the complex phasor approach, the aberrated pupil function is defined as

$$P(\hat{u}_0,\hat{v}_0;x,y) = \text{circ}\left(\frac{\sqrt{x^2+y^2}}{w_{XP}}\right)\exp\left[-jkW\left(\hat{u}_0,\hat{v}_0;\frac{x}{w_{XP}},\frac{y}{w_{XP}}\right)\right], \quad (8.3)$$

where the circle function describes the diffraction-limited XP discussed in Chapter 7. For the Seidel polynomial W, an image plane coordinate is required, and the pupil coordinates are normalized by the XP radius w_{XP}.

Referring to Fig. 8.1, W is commonly assigned a positive value when W_{ab} lies to the image side of the ideal spherical wavefront W_{sp}.[1] This is opposite to the phasor phase-front convention where the leading phase is more negative, so a minus sign is included in the exponent of Eq. (8.3).

8.3.2 Imaging transfer functions

The coherent and optical transfer function definitions of Eqs. (7.7) and (7.26) still apply for an aberrated imaging system. These are repeated here with the normalized image coordinates included. The coherent transfer function is

$$H(\hat{u}_0,\hat{v}_0;f_U,f_V) = P(\hat{u}_0,\hat{v}_0;-\lambda z_{XP}f_U,-\lambda z_{XP}f_V), \quad (8.4)$$

and the optical transfer function (OTF) is

$$\mathcal{H}(\hat{u}_0,\hat{v}_0;f_U,f_V) = H(\hat{u}_0,\hat{v}_0;f_U,f_V) \star H(\hat{u}_0,\hat{v}_0;f_U,f_V)\big|_{\text{norm}}. \quad (8.5)$$

As discussed in Section 7.3.1, the normalization makes the OTF have a value of unity at DC. Our focus here is incoherent imaging. Even though the aberrated OTF can be significantly different than the diffraction-limited OTF, it can be shown that the incoherent cutoff frequency remains the same ($2f_0$).[5] For a computer simulation of an aberrated system, the fundamental sampling restriction remains

$$2f_o \leq f_N, \quad (8.6)$$

where $2f_0 = 2w_{XP}/(\lambda z_{XP}) = 1/\left[\lambda(f\#)\right]$ and $f_N = 1/(2\Delta u)$. The sample interval in the image plane is Δu.

8.4 Image Quality

When studying imaging system performance, full-image simulations are not often done. Instead, image quality metrics are used to predict image performance. Although a full-image simulation is demonstrated in Section 8.7, our first task is to discuss and demonstrate the PSF and the MTF, two common image quality metrics.

8.4.1 Point spread function

The PSF (incoherent impulse response $|h|^2$) indicates the characteristics of the image of a point source. Roughly speaking, the wider the PSF, the poorer the imaging resolution. A direct approach to obtain the PSF in the computer is

$$\left|h(\hat{u}_0,\hat{v}_0;u,v)\right|^2 = \left|\mathfrak{I}^{-1}\{H(\hat{u}_0,\hat{v}_0;f_U,f_V)\}\right|^2. \tag{8.7}$$

It is explicit in Eq. (8.7) that the PSF is dependent on the image position coordinate ($\hat{u}_0, \hat{v}_0$). This means the aberrated imaging system is space variant, where a different impulse response (PSF) is required for every position in the image.

8.4.2 Modulation transfer function

For diffraction-limited systems, the OTF $\mathcal{H}$ is a real function, but if the pupil function is complex, the OTF can be complex. The OTF describes both the attenuation and phase that are impressed upon the spatial frequencies of an ideal incoherent image as formed by the imaging system. The effect of a phase change for a given spatial frequency can be thought of as translation of the corresponding sinusoidal spatial component in the image plane.

The modulus of the OTF $|\mathcal{H}|$ is known as the *modulation transfer function*. The MTF simply describes attenuation of the sinusoidal image irradiance components as a function of spatial frequency. The terms "contrast" and "modulation depth" are also applied to MTF values. The modulation value for a single-frequency sinusoidal irradiance pattern can be calculated from measurements using

$$\text{Modulation} = \frac{I_{MAX} - I_{MIN}}{I_{MAX} + I_{MIN}}, \tag{8.8}$$

where I_{MAX} is the irradiance measured at a peak of the pattern, and I_{MIN} is the irradiance measured at a valley. A modulation value of 1 indicates no degradation of the sinusoidal spatial frequency component. A value of zero means the spatial frequency cannot be produced by the imaging system.

The MTF tells much of the story about the spatial response of an imaging system and is easier to measure on real systems than the OTF. Once the PSF is in hand, the MTF can be found by

$$\text{MTF}(\hat{u}_0,\hat{v}_0;f_U,f_V) = \left|\mathfrak{I}\left\{\left|h(\hat{u}_0,\hat{v}_0;u,v)\right|^2\right\}\right|\bigg|_{norm}. \tag{8.9}$$

8.5 Lens Example—PSF and MTF

Figure 8.4 shows a ray trace layout diagram for an actual plano-convex lens generated by the commercial optical design software ZEMAX. ZEMAX is used here to obtain some practical aberration values for the lens simulation. The effective focal length of the lens is 100 mm and the pupil has a diameter of 20 mm, which gives an image space *f*-number of 5. It is assumed that the lens will image a distant object; therefore, $z_{XP} \approx f$. The actual prescription data for the

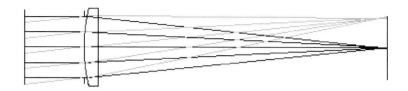

Figure 8.4 ZEMAX ray trace layout for $f/5$ plano-convex lens. Focal length (effective) = 100 mm, pupil diameter = 20 mm. Image positions are shown for 0 and 10 mm.

lens is a front surface radius of curvature = 51.68 mm, a back surface radius of curvature = infinite, glass = BK7, and center thickness = 4.585 mm.

The maximum image height is arbitrarily chosen to be 10 mm. In Fig. 8.4, this is the point on the upper part of the image plane (y axis) to which the off-axis ray bundle is converging. The normalized image coordinate for this point is $\hat{u}_0 = 0, \hat{v}_0 = 1$. The field angle (the angle the central ray in the bundle makes with the optical axis) corresponding to this point is 5.72 deg. The maximum image height is associated with the maximum *field of view* for the system.

The primary Seidel aberration coefficients as calculated by ZEMAX for this lens arrangement are listed in Table 8.2. The wavelength assumed is 0.55 μm. The coefficient values are listed in waves, so λ is a multiplier to get the path length value. The amount of aberration is considerable; five waves of spherical causes significant spreading of the PSF and nine waves of astigmatism and seven waves of curvature create considerably more image degradation for off-axis image points.

The script "lens_psfmtf" computes and displays the PSF and MTF for the $f/5$ lens:

```
1   % lens_psfmtf
2   % f/5 plano-convex lens (Newport lens KPX094)
3   % 10 mm image height
```

Table 8.2 Seidel coefficient values for $f/5$ lens (from ZEMAX).

Coefficient	Value[a]
W_d	0
W_{040}	4.963λ
W_{131}	2.637λ
W_{222}	9.025λ
W_{220}[b]	7.536λ
W_{311}	0.157λ

[a] $\lambda = 0.55$ μm.
[b] Sagittal field curvature.

```
4   % psf and mtf with Seidel aberrations
5   % aberration coefficients from ZEMAX
6   M=1024;              %sample #
7   L=1e-3;              %image plane side length
8   du=L/M;              %sample interval
9   u=-L/2:du:L/2-du; v=u; %coordinates
10
11  lambda=0.55e-6;      %wavelength
12  k=2*pi/lambda;       %wavenumber
13  Dxp=20e-3; wxp=Dxp/2; %exit pupil size
14  zxp=100e-3;          %exit pupil distance
15  fnum=zxp/(2*wxp);    %exit pupil f-number
16  lz=lambda*zxp;
17  twof0=1/(lambda*fnum);%incoh cutoff freq
18
19  u0=0; v0=0;          %normalized image coordinate
20
21  % aberration coefficients
22  wd=0*lambda;
23  w040=4.963*lambda;
24  w131=2.637*lambda;
25  w222=9.025*lambda;
26  w220=7.536*lambda;
27  w311=0.157*lambda;
28
29  fu=-1/(2*du):1/L:1/(2*du)-(1/L); %image freq coords
30  [Fu,Fv]=meshgrid(fu,fu);
31
32  % wavefront
33  W=seidel_5(u0,v0,-lz*Fu/wxp,-lz*Fv/wxp,...
34      wd,w040,w131,w222,w220,w311);
35
36  % coherent transfer function
37  H=circ(sqrt(Fu.^2+Fv.^2)*lz/wxp).*exp(-j*k*W);
38  figure(1);
39  imagesc(u,v,angle(H)); axis xy; axis square
40  xlabel('u (m)'); ylabel('v (m)'); colormap('gray')
41
42  % point spread function
43  h2=abs(ifftshift(ifft2(fftshift(H)))).^2;
44
45  figure(2)       % psf image and profiles
46  imagesc(u,v,nthroot(h2,2)); axis xy; axis square
47  xlabel('u (m)'); ylabel('v (m)'); colormap('gray')
48  figure(3);
49  plot(u,h2(M/2+1,:)); xlabel('u (m)'); ylabel('PSF');
50  figure(4);
51  plot(u,h2(:,M/2+1)); xlabel('v (m)'); ylabel('PSF');
52
53  % MTF
```

```
54  MTF=fft2(fftshift(h2));
55  MTF=abs(MTF/MTF(1,1));   %normalize DC to 1
56  MTF=ifftshift(MTF);
57
58  % analytic MTF
59  MTF_an=(2/pi)*(acos(fu/twof0)-(fu/twof0)...
60      .*sqrt(1-(fu/twof0).^2));
61  MTF_an=MTF_an.*rect(fu/(2*twof0));  %zero after cutoff
62
63  figure(5)       % MTF profiles
64  plot(fu,MTF(M/2+1,:),fu,MTF(:,M/2+1),':',...
65      fu,MTF_an,'--');
66  axis([0 150000 0 1]);
67  legend('u MTF','v MTF','diff limit');
68  xlabel('f (cyc/m)'); ylabel('Modulation');
```

Some comments on this script are as follows:

(a) Line 19: Input line for the normalized image position of interest.

(b) Line 21: Aberration coefficients taken from Table 8.1.

(c) Line 39: It is a good idea to check the resulting coherent transfer function. If the diameter is half the size or less than the array, then the sampling criterion of Eq. (8.6) is satisfied.

(d) Line 55: The MTF is normalized to 1 at the DC frequency value.

(e) Line 58: The 1D analytic diffraction-limited MTF [from Eq. (7.31)] is generated for comparison with the aberrated results.

The axial image point ($\hat{u}_0 = 0, \hat{v}_0 = 0$) is selected in line 19, and executing the script produces the results shown in Fig. 8.5. The phase of the coherent transfer function in Fig. 8.5(a) is displayed as a kind of contour map, where black indicates zero and white is 2π rad. The phase is actually sampled adequately, but some patterning is added to the image during display. The idea is to see if the diameter of the transfer function is half the size or less than the array, so the criterion of Eq. (8.6) is satisfied. The PSF in Figs. 8.5(b)–(d) appears to be relatively narrow—which is good—but, in fact, it is spread considerably relative to the diffraction-limited PSF (see Exercise 8.8). Only one half of the MTF curve is usually displayed since the negative frequency side is the same. Figure 8.5(e) shows the MTF drops quickly with increasing spatial frequency compared to the diffraction-limited curve. A practical cutoff frequency estimate from the MTF curve is perhaps 120 or 130 cyc/mm. For the axial image point, the PSF and the MTF are symmetric, so there is no difference between the u- and v-axis results.

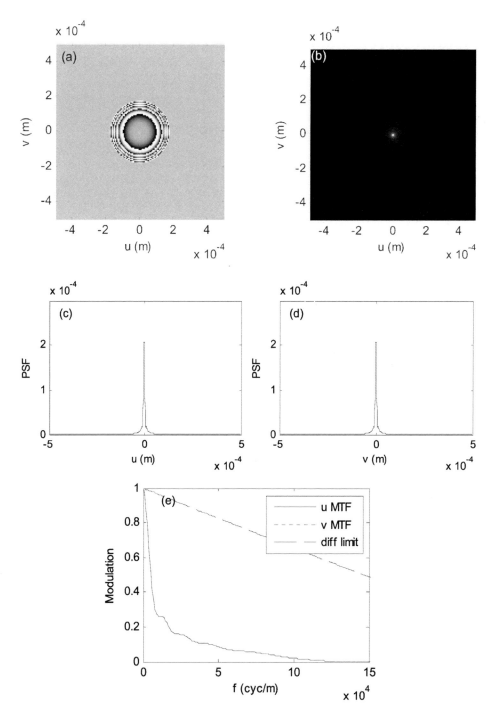

Figure 8.5 Plano-convex lens results for the axial image point $\hat{u}_0 = 0, \hat{v}_0 = 1$: (a) coherent transfer function phase; (b) PSF image; (c) PSF u-axis profile; (d) PSF v-axis profile; and (e) MTF. Some of the patterning in (a) is display artifact.

Change the maximum height image point on the v-axis ($\hat{u}_0 = 0, \hat{v}_0 = 1$) and the results are shown in Fig. 8.6. The PSF, no longer symmetric, has a characteristic elongation in the v direction [Fig. 8.6(b)–(d)]. The peak irradiance value is much smaller than the value in Fig. 8.5 due to considerable spreading of the PSF. The MTFs [Fig. 8.6(e)] are different along the two axes, with the v-axis MTF fairing the worst. Practical cutoff frequencies are approximately 20 cycles/mm for u and perhaps 10 cycles/mm for v.

Keep in mind that although the PSF results are centered in arrays corresponding to the 1 mm × 1 mm area, in terms of the full image plane the axial PSF is positioned on the optical axis, whereas the maximum image height PSF is found to be 10 mm off the optical axis.

For verification, or at least a sanity check, these results can be compared with analysis performed by ZEMAX for the same example. The ray spot diagrams (Fig. 8.7) are created by launching a set of rays in the pupil and tracing the rays in a rectilinear fashion to the image plane. The ray locations are marked in the image plane. The side length of the ZEMAX plots is the same as the `lens_psfmtf` results. The ray spot diagrams resemble the shapes of the PSFs generated by the `lens_psfmtf` script, which is a good sign. A more specific comparison can be made with the MTFs (Fig. 8.8). "T" refers to *tangential plane*, which is a plane aligned with our v axis. "S" refers to *sagittal plane*, which for our purposes includes the u axis. It is difficult to identify the T and S curves for the 10 mm image point, but the T (v axis) is the lower curve. The ZEMAX curves appear quite similar to the `lens_psfmtf` curves. Some differences are expected between the two sets of results because `lens_psfmtf` is working with only the five primary Seidel aberrations, whereas ZEMAX includes higher-order terms in the pupil wavefront description.

8.6 Wavefront Sampling

The criterion in Eq. (8.6) is associated with the incoherent cutoff frequency, but a simulation can also give erroneous results when the phase of the wavefront OPD in the pupil becomes undersampled. If the slope of the wavefront OPD is large, such that there is more than a π radian change in the phase quantity kW over the sample interval, then the pupil phase will be aliased. This criterion can be written as

$$\Delta \hat{x} k \left| \frac{\partial W(\hat{x}, \hat{y})}{\partial \hat{x}} \right|_{max} \leq \pi, \quad \text{and} \quad \Delta \hat{y} k \left| \frac{\partial W(\hat{x}, \hat{y})}{\partial \hat{y}} \right|_{max} \leq \pi. \qquad (8.10)$$

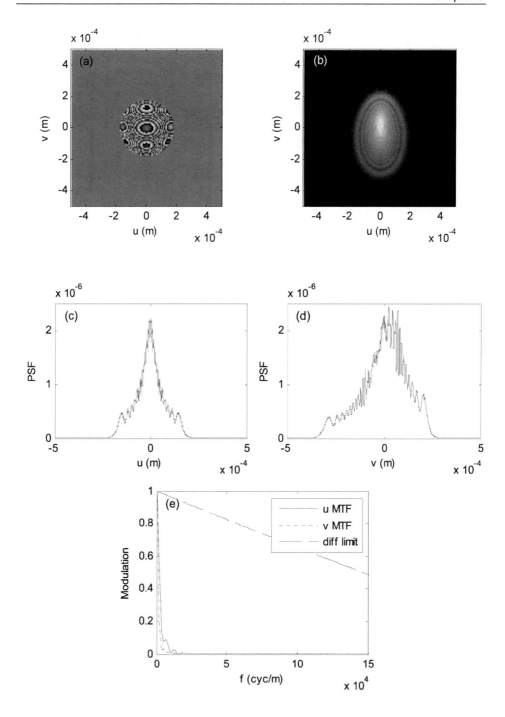

Figure 8.6 Plano-convex lens results for the maximum field image point $\hat{u}_0 = 0$, $\hat{v}_0 = 1$: (a) coherent transfer function phase; (b) PSF image; (c) PSF u-axis profile; (d) PSF v-axis profile; and (e) MTF. Some of the patterning in (a) is display artifact.

Wavefront Aberrations

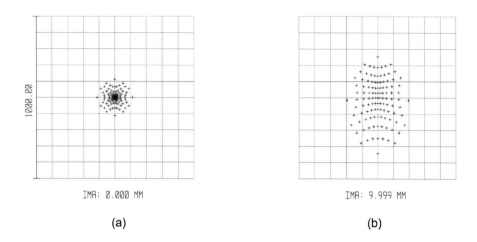

Figure 8.7 ZEMAX ray trace spot diagrams for the $f/5$ plano-convex lens example. (a) axial image PSF and (b) maximum image height (v axis) PSF.

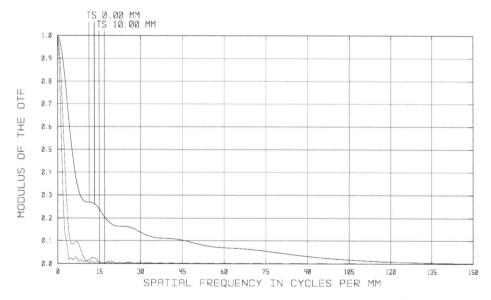

Figure 8.8 ZEMAX MTF curves for the $f/5$ plano-convex lens example. The dark curve corresponds to the on-axis image point and the light curves are the tangential (T) and sagittal (S) planes for the $\hat{u}_0 = 0$, $\hat{v}_0 = 1$ image point.

In general, $W(\hat{x}, \hat{y})$ is not symmetric, so both axes may need consideration. For one axis, the normalized pupil sample interval is related to the XP and image plane parameters by

$$\Delta \hat{x} = \frac{\lambda z_{XP}/L}{w_{XP}} = \frac{2\lambda(f/\#)}{L}. \tag{8.11}$$

Recall that L is the image plane side length. Substituting Eq. (8.11) into Eq. (8.10) and rearranging yields

$$\left|\frac{\partial W(\hat{x}, \hat{y})}{\partial \hat{x}}\right|_{max} \leq \frac{L}{4(f/\#)}. \tag{8.12}$$

For illustration, consider the primary Seidel aberration polynomial. The x axis will usually produce the largest wavefront slopes, so

$$\frac{\partial W(\hat{x}, \hat{y})}{\partial \hat{x}} = 2W_d \hat{x} + 4W_{040}\left(\hat{x}^2 + \hat{y}^2\right)\hat{x}$$
$$+ W_{131}\hat{u}_0\left(3\hat{x}^2 + \hat{y}^2\right) + 2W_{222}\hat{u}_0^2 \hat{x} \tag{8.13}$$
$$+ 2W_{220}\hat{u}_0^2 \hat{x} + W_{311}\hat{u}_0^3.$$

The maximum value of Eq. (8.13) depends on the signs and values of the coefficients and coordinates, but a few special cases can be examined.

For an axial image point, $\hat{u}_0 = 0$ and assuming no defocus, the maximum occurs when $\left(\hat{x}^2 + \hat{y}^2\right)\hat{x} = 1$, which yields

$$\left|\frac{\partial W(\hat{x}, \hat{y})}{\partial \hat{x}}\right|_{max} = 4W_{040}. \tag{8.14}$$

For example, consider the axial image point of the plano-convex lens example of Section 8.5. Combining Eqs. (8.12) and (8.14) gives

$$W_{040} \leq \frac{L}{16(f/\#)}. \tag{8.15}$$

With $L = 1 \times 10^{-3}$ m and $f/\# = 5$, the criteria is $W_{040} \leq 12.5 \times 10^{-6}$ m, or in terms of waves, $W_{040} \leq 22.73\lambda$. The plano-convex lens spherical aberration (Table 8.2) is well within this bound.

Another example is when all the aberration coefficients are positive, such as the case for the example lens (Table 8.2), then

$$\left|\frac{\partial W(\hat{x},\hat{y})}{\partial \hat{x}}\right|_{max} = 2W_d + 4W_{040} + 3W_{131}\hat{u}_0 + 2W_{222}\hat{u}_0^2 + 2W_{220}\hat{u}_0^2 + W_{311}\hat{u}_0^3, \qquad (8.16)$$

where the maximum requires $\hat{x} = 1$ (so $\hat{y} = 0$) (see Exercise 8.5).

8.7 Superposition Imaging Example

In the `lens_psfmtf` example, a 1024×1024 sample array is used to examine a 1×1 mm area of the image plane. To model the full image plane (radius of 10 mm) with the same sampling would require roughly $20{,}000 \times 20{,}000$ samples—which is not practical. Furthermore, since the aberrated imaging system is space variant, a simple convolution approach cannot be used. The resulting superposition operation leads to long computer run times. This is all to say that it can be impractical to directly model the full-image field of practical imaging systems. However, it is still a good exercise to see how the impulse functions can be distributed in an image frame and to learn to set up a superposition problem. To do this we need to conjure up a problem that can be handled with fewer samples.

8.7.1 Image plane PSF map

Assume an $f/20$ XP and $\lambda = 0.5$ μm. This results in $2f_0 = 10^5$ m^{-1}. Working with $M = 250$ and an image plane side length of $L = 1$ mm leads to $f_N = 1.25 \times 10^5$ m^{-1}. The criterion of Eq. (8.6) is thus satisfied. For an $f/20$ system, the PSF would, typically, change very little as a function of position within the 1×1 mm image frame. But, to illustrate how the PSF might change, the amount of aberration is exaggerated.

The following script, `psf_map` generates a map of PSFs for different positions in the image plane. The map helps illustrate the overall effect of the system response. The example assumes ½ wave of spherical aberration, 1 wave of coma, and 1.5 waves of astigmatism. The concept is to loop through various image plane coordinates, creating PSFs for each, and the shifting the PSFs to the appropriate position in an image plane array. Remember that the PSFs are all produced initially in the center of the array.

```
1  % psf_map generate psf map
2
3  M=250;
4  L=1e-3;                %image plane side length
5  du=L/M;                %sample interval
6  u=-L/2:du:L/2-du; v=u; %coordinates
7
8  lambda=0.5*10^-6;      %wavelength
```

```
9   k=2*pi/lambda;              %wavenumber
10  wxp=2.5e-3;                 %exit pupil radius
11  zxp=100e-3;                 %exit pupil distance
12  fnum=zxp/(2*wxp)            %exit pupil f-number
13
14  twof0=1/(lambda*fnum)       %inc cutoff freq
15  fN=1/(2*du)                 %Nyquist frequency
16
17  % aberration coefficients
18  wd=0*lambda;
19  w040=0.5*lambda;
20  w131=1*lambda;
21  w222=1.5*lambda;
22  w220=0*lambda;
23  w311=0*lambda;
24
25  fu=-1/(2*du):1/L:1/(2*du)-(1/L); %image freq coords
26  fu=fftshift(fu); %shift cords, avoid shift H in loop
27  [Fu,Fv]=meshgrid(fu,fu);
28
29  I=zeros(M);
30  % loop through image plane positions
31  for u0=[-.7:.7/3:.7]
32      for v0=[-.7:.7/3:.7]
33          % wavefront
34          W=seidel_5(u0,v0,-2*lambda*fnum*Fu...
35              ,-2*lambda*fnum*Fv,...
36              wd,w040,w131,w222,w220,w311);
37          % coherent transfer function
38          H=circ(sqrt(Fu.^2+Fv.^2)*2*lambda*fnum)...
39              .*exp(-j*k*W);
40          % PSF
41          h2=abs(ifftshift(ifft2(H))).^2;
42          % shift PSF to image plane position
43          h2=circshift(h2,[round(v0*M/2)...
44              ,round(u0*M/2)]);
45          % add into combined frame
46          I=I+h2;
47      end
48  end
49
50  figure(1)
51  imagesc(u,v,nthroot(I,2));
52  xlabel('u (m)'); ylabel('v (m)');
53  colormap('gray'); axis square; axis xy
```

Some comments on `psf_map`:

(a) Line 14: Semicolons are left off of the incoherent cutoff ($2f_0$) and Nyquist frequency (f_N) to help check the sampling when changing parameters.

(b) Line 26: An execution time-saving feature. The frequency coordinate array `fu` is shifted, so `H` can be computed in the loop in the shifted position—thus avoiding the `fftshift` of `H` before the `ifft2` in line 41.

(c) Line 29: Output irradiance array `I` is dimensioned.

(d) Lines 31 and 32: Seven `u0` and `v0` positions selected.

(e) Line 43: After the PSF is computed for a particular `u0`, `v0` coordinate, the PSF is shifted to the `u0`, `v0` coordinate in the image plane. The function `circshift` is a quick way to accomplish the shift. `circshift` shifts array elements along the rows and columns. Elements that shift off one side reappear on the other side of the array—similar to the periodic extension concept. This is an indexing operation, and since MATLAB organizes arrays in row/column order, the first variable in `circshift` corresponds to the y axis (`v0`). Round forces the row/column input values to be integers. The image plane width is assumed to be just wide enough to accommodate the u0=±1 or v0=±1 image points.

(f) Line 46: The PSFs are simply added together in the array `I`. Don't mistake this for a convolution or superposition operation. It is just an arrangement to quickly view and compare the PSFs.

The PSF image plane map result from `psf_pt_array` is shown in Fig. 8.9. The center PSF is symmetric, but toward the image plane edges the PSFs become elongated due to coma and astigmatism.

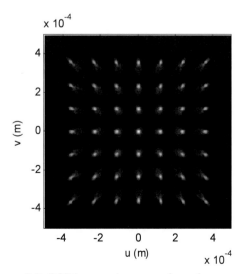

Figure 8.9 PSF image plane map for psf_map example.

8.7.2 Image simulation

Space-variant incoherent imaging cannot be described by the convolution in Eq. (7.23). Instead, the superposition integral is needed,

$$I_i(\hat{u}_0,\hat{v}_0) = \int\int_{-\infty}^{\infty} I_g(u,v)|h(\hat{u}_0,\hat{v}_0;u,v)|^2 \, du\, dv, \qquad (8.17)$$

where the image is defined in normalized coordinates $\hat{u}_0, \hat{v}_0$. The PSF $|h(\hat{u}_0,\hat{v}_0;u,v)|^2$ is defined in Eq. (8.7). The approach for image simulation is as follows:

(1) Select a point $\hat{u}_0, \hat{v}_0$ in the image plane.
(2) Generate the PSF corresponding to $\hat{u}_0, \hat{v}_0$. Normalize the PSF to a volume of 1 (this is equivalent to the OTF or MTF normalization).
(3) Move the PSF to the $\hat{u}_0, \hat{v}_0$ image position in the array. Multiply the PSF by the ideal image value at $\hat{u}_0, \hat{v}_0$.
(4) Add the PSF result into the image array.
(5) Return to step 1 and perform computations for the next image coordinate.

The following script, `image_super`, is a superposition imaging simulation that applies the aberration parameters for the `psf_map` example to the test chart ideal image:

```
1   % image_super superposition image
2
3   A=imread('USAF1951B250','png');
4   [M,N]=size(A); A=flipud(A);
5   Ig=single(A); Ig=Ig/max(max(Ig));
6
7   L=1e-3;                 %image plane side length
8   du=L/M;                 %sample interval
9   u=-L/2:du:L/2-du; v=u;  %coordinates
10  fN=1/(2*du)             %Nyquist frequency
11
12  lambda=0.5*10^-6;       %wavelength
13  k=2*pi/lambda;          %wavenumber
14  wxp=2.5e-3;             %exit pupil radius
15  zxp=100e-3;             %exit pupil distance
16  fnum=zxp/(2*wxp)        %exit pupil f-number
17
18  twof0=1/(lambda*fnum)   %inc cutoff freq
19  fN=1/(2*du)             %Nyquist frequency
```

Wavefront Aberrations

```
20
21  % aberration coefficients
22  wd=0*lambda;
23  w040=0.5*lambda;
24  w131=1*lambda;
25  w222=1.5*lambda;
26  w220=0*lambda;
27  w311=0*lambda;
28
29  % frequency coordinates
30  fu=-1/(2*du):1/L:1/(2*du)-(1/L);
31  fu=fftshift(fu); %shift cords, avoid shift H in loop
32  [Fu,Fv]=meshgrid(fu,fu);
33
34  I=zeros(M);
35  % loop through image plane positions
36  for n=1:M
37      v0=(n-(M/2+1))/(M/2) %norm v image coord
38      for m=1:M
39          u0=(m-(M/2+1))/(M/2); %norm u image coord
40          % wavefront
41          W=seidel_5(u0,v0,-2*lambda*fnum*Fu,...
42              -2*lambda*fnum*Fv,...
43              wd,w040,w131,w222,w220,w311);
44          % coherent transfer function
45          H=circ(sqrt(Fu.^2+Fv.^2)*2*lambda*fnum)...
46              .*exp(-j*k*W);
47          % create PSF - normalize volume to 1
48          h2=abs(ifftshift(ifft2(H))).^2;
49          h2=h2/(sum(sum(h2)));
50          % shift h2 to image plane position
51          h2=circshift(h2,[n-(M/2+1),m-(M/2+1)]);
52          % superposition
53          I=Ig(n,m)*h2+I;
54      end
55  end
56
57  figure(1)
58  imagesc(u,v,nthroot(I,3));
59  colormap('gray'); axis square; axis xy
60  xlabel('x (m)'); ylabel('y (m)');
```

Some comments on `image_super`:

(a) Line 31: `fu` is shifted to avoid using fftshift on `H` in line 48 (a time saver).

(b) Line 35: `for` loops are used to step through each position in the image plane corresponding to all coordinate pairs $\hat{u}_0, \hat{v}_0$.

(c) Lines 36 and 37: The first loop involves the row index n. Because of MATLAB's row/column ordering, this index corresponds to the *y*-axis variable v0. This is one of the few times in this book where it is necessary to use the index values to derive physical parameter values. Looping with the index is forced by line 53, where each ideal image element is addressed.

(d) Lines 38 and 39: The second loop involves the row index m corresponding to the *x*-axis variable u0.

(e) Line 49: The PSF is normalized to a volume of 1, equivalent to the OTF normalization in Eq. 7.25.

(f) Line 51: circshift is applied using the index variables, which has the same effect as the approach used with circshift in psf_map.

(g) Line 53: Finally, the superposition [Eq. (8.17)] is performed.

(h) A final comment: This code takes a relatively long time to execute! (About an hour on my laptop.) Individual PSFs ($250 \times 250 = 62{,}500$) are created and incorporated in the superposition frame. Compare this with a convolution result that involves a single PSF and can be accomplished in fractions of a second with a few fast Fourier tranforms. The luxury of a convolution is not available in this case.

The result for image_super is shown in Fig. 8.10. The off-axis aberration effects are obvious in the edges of the image. The overall image appearance is consistent with the corresponding PSF map displayed in Fig. 8.9. This illustrates the usefulness of the PSF for predicting image results.

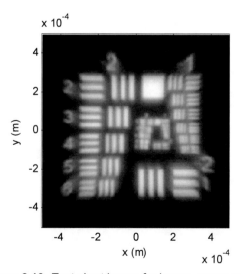

Figure 8.10 Test chart image for image_super example.

Try this example. You can always get a bite to eat, watch a movie, do some more homework...while you are waiting for the result.

8.7.3 Practical image simulation

What can be done to address the issues in the simulation approach of Section 8.7.2 (namely, that the code takes a long time to execute and the simulation image plane area is small)? Well, get a faster computer with more memory. However, there are other more clever ways to address these issues. The execution time can be reduced significantly with some code changes. For example, the PSFs in each quadrant of the image are all the same—just rotated—so, reusing rather than recalculating the PSFs saves time.

The bigger problem is the small image plane area due to the limited array size. An option is to undersample the image significantly—spread out the samples—and generate approximate PSFs that are also undersampled. Depending on the ideal image, this may give a useful approximation of the full-image result.

An approach to address both the area and time problem is to divide the image plane into sections (tiles). A representative PSF for each tile is assumed and convolution processing is applied to each tile. The resulting tiles are reassembled to create the full image. The result is approximate since the PSF does not vary continually over the image plane. Also, artifacts are likely to be present in the reassembled image having to do with "stitching" the tiles together. Nevertheless, this can be an effective approach when short execution time and a modest image field are important.

8.8 Exercises

8.1 For the $f/5$ plano-convex lens example in Section 8.5:

(a) What is the incoherent cutoff frequency? What is the Nyquist frequency? Is Eq. (8.6) satisfied?

(b) Generate a surface plot of the wavefront OPD for the on-axis point $(\hat{u}_0, \hat{v}_0) = (0,0)$ and the maximum fractional image point $(\hat{u}_0, \hat{v}_0) = (0,1)$.

8.2 Spherical aberration arises from the use of spherical optical surfaces to focus light. Unfortunately, a spherical surface does not produce a spherical converging wave for most incident wavefronts. Focus, or more precisely, *defocus*, can be applied to counter the effect of spherical aberration.

(a) Assume an axial image point $(\hat{u}_0, \hat{v}_0) = (0,0)$, $W_{040} = 1$ (chosen arbitrarily) and zero for the other Seidel aberration coefficients. Create a surface plot of W.

(b) Include some defocus W_d in (a). Examine the surface plot of W. Adjust W_d to "compensate" for W_{040}. Remember, ideally W should be flat. Make a rough judgment on the amount of W_d needed for the best compensation.

8.3 PSF through focus: A useful optical design technique is to examine the PSF of a system as a function of focus. The center of curvature of the defocus wavefront reference sphere is positioned a distance $\delta_d = 8(f/\#)^2 W_d$ from the original image point.[2] For example, with an $f/10$ system, where $W_d = -1.0\lambda$ and $\lambda = 0.5$ μm, then $\delta_d = -0.4$ mm. Thus, if the image plane is moved -0.4 mm toward the lens system, the effect is as if an aberration of $W_d = -1.0\lambda$ is created in the pupil.

For the $f/5$ lens example in Section 8.5, examine the PSFs for the axial image point as a function of focus where $\delta_d = -1, -0.5, 0, +0.5,$ and $+1$ mm. Which value of δ_d appears to create the "narrowest" PSF (or the PSF with the highest peak value)? What is the corresponding value of W_d?

8.4 Parabolic Mirror: A mirror with a parabolic curvature has a diffraction-limited PSF for an incident plane wave at zero field angle (traveling down the optical axis). Figure 8.10 shows a ray trace diagram from ZEMAX of an $f/5$ parabolic mirror arrangement. In practice, the converging light is usually directed out of the incoming beam with a second mirror, for example, a small flat mirror at a 45-deg angle, but that issue is ignored here. The mirror parameters are $f = 200$ mm, diameter = 40 mm, and the maximum image height is 3.5 mm. The Seidel aberration coefficients are shown in Table 8.3 for the He–Ne laser wavelength 0.633 μm (a common wavelength for component testing).

Alter the `lens_psfmtf` script to model this mirror. Use $M = 1024$ and $L = 0.1 \times 10^{-3}$ m. Generate the PSF and MTF for the following image position coordinates $(\hat{u}_0, \hat{v}_0)$: (0, 0), (0, 1), and (0.707, 0.707).

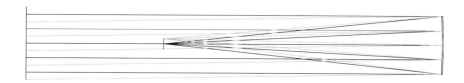

Figure 8.10 ZEMAX ray trace layout for $f/5$ parabolic mirror. Focal length (effective) = 200 mm, pupil diameter = 40 mm. Incident field angles are 0 and 1 deg, corresponding to image plane heights of 0 and 3.5 mm.

Table 8.3 Seidel coefficient values for $f/5$ parabolic mirror (ZEMAX).

Coefficient	Value[a]
W_d	0
W_{040}	0
W_{131}	-1.3792λ
W_{222}	0.4815λ
W_{220}[b]	0
W_{311}	0

[a] $\lambda = 0.633$ μm.
[b] Sagittal field curvature.

8.5 For the maximum field image point ($\hat{u}_0 = 1$) in the $f/5$ lens example of Section 8.5, find the amount of positive defocus W_d that can be added to the pupil such that Eq. (8.12) is an equality (critically satisfied). Apply defocus in the `lens_psfmtf` script and examine the PSF for defocus values over and under the critical value. What happens to the PSF when the wavefront OPD is undersampled?

8.6 Zernike Polynomials: Zernike polynomials are another common method of describing the wavefront OPD of a pupil.[1] They are a set of terms for a circular pupil that are orthogonal (changing one term does not affect the other). Unlike the Seidel series, Zernike polynomials do not assume circular symmetry. There is no requirement to rotate the coordinates of the Zernike system for different image points. When measuring an unknown wavefront with an instrument such as an interferometer, the usual approach for characterizing the wavefront is to "fit" the measurements with Zernike terms. This means finding term coefficients that result in the closest match to the wavefront. Zernike polynomials are also used to characterize optical wavefront distortions due to other phenomena such as atmospheric turbulence.[6,7]

Table 8.4 Zernike polynomials for f/5 plano-convex lens at $\hat{u}_0 = 0, \hat{v}_0 = 1$.

Number	Coefficients	Polynomial
1	7.70185740	1
2	0.00000000	4^(1/2)(p) * COS (A)
3	1.04836665	4^(1/2)(p) * SIN (A)
4	4.95277338	3^(1/2)(2p^2 - 1)
5	0.00000000	6^(1/2)(p^2) * SIN (2A)
6	-1.83668298	6^(1/2)(p^2) * COS (2A)

7	0.38477648	8^(1/2)(3p^3 - 2p) * SIN (A)
8	0.00000000	8^(1/2)(3p^3 - 2p) * COS (A)
9	-0.00829929	8^(1/2)(p^3) * SIN (3A)
10	0.00000000	8^(1/2)(p^3) * COS (3A)
11	0.39476169	5^(1/2)(6p^4 - 6p^2 + 1)
12	-0.00582307	10^(1/2)(4p^4 - 3p^2) * COS (2A)
13	0.00000000	10^(1/2)(4p^4 - 3p^2) * SIN (2A)
14	0.00007007	10^(1/2)(p^4) * COS (4A)
15	0.00000000	10^(1/2)(p^4) * SIN (4A)
16	0.00000000	12^(1/2)(10p^5 - 12p^3 + 3p) * COS (A)
17	0.00782021	12^(1/2)(10p^5 - 12p^3 + 3p) * SIN (A)
18	0.00000000	12^(1/2)(5p^5 - 4p^3) * COS (3A)
19	-0.00010086	12^(1/2)(5p^5 - 4p^3) * SIN (3A)
20	0.00000000	12^(1/2)(p^5) * COS (5A)
21	0.00000048	12^(1/2)(p^5) * SIN (5A)
22	0.00233053	7^(1/2)(20p^6 - 30p^4 + 12p^2 - 1)
23	0.00000000	14^(1/2)(15p^6 - 20p^4 + 6p^2) * SIN (2A)
24	-0.00010805	14^(1/2)(15p^6 - 20p^4 + 6p^2) * COS (2A)
25	0.00000000	14^(1/2)(6p^6 - 5p^4) * SIN (4A)
26	0.00000126	14^(1/2)(6p^6 - 5p^4) * COS (4A)
27	0.00000000	14^(1/2)(p^6) * SIN (6A)
28	0.00000006	14^(1/2)(p^6) * COS (6A)
29	0.00008738	16^(1/2)(35p^7 - 60p^5 + 30p^3 - 4p) * SIN (A)
30	0.00000000	16^(1/2)(35p^7 - 60p^5 + 30p^3 - 4p) * COS (A)
31	-0.00000163	16^(1/2)(21p^7 - 30p^5 + 10p^3) * SIN (3A)
32	0.00000000	16^(1/2)(21p^7 - 30p^5 + 10p^3) * COS (3A)
33	0.00000000	16^(1/2)(7p^7 - 6p^5) * SIN (5A)
34	0.00000000	16^(1/2)(7p^7 - 6p^5) * COS (5A)
35	0.00000002	16^(1/2)(p^7) * SIN (7A)
36	0.00000000	16^(1/2)(p^7) * COS (7A)
37	0.00001672	9^(1/2)(70p^8 - 140p^6 + 90p^4 - 20p^2 + 1)

Table 8.4 is a ZEMAX output listing of the Zernike coefficients ("standard coefficients") for the $f/5$ plano-convex lens example. The listing corresponds to the image point $\hat{u}_0 = 0, \hat{v}_0 = 1$. The specific polynomial forms are also listed as there are several versions of Zernike polynomials in the literature. p is the normalized radial distance in the pupil and A is the polar angle referenced to the *x*

axis. Term 1 is a "piston" term that has no effect on the PSF or MTF calculation. Note that the coefficients after 11 are small in this case.

(a) Write a function like seidel_5 but to evaluate the Zernike polynomials. Include terms through at least No. 11. The pupil coordinates are as follows:

```
p=sqrt(X.^2+Y.^2)
A=atan2(Y,X)
```

In the Zernike case, the image point position is not an explicit part of the polynomial description. There are various efficient ways to code up Zernike polynomials, for example, see Schmidt,[6] but there is always the straightforward way of listing them out as was done for seidel_5.

(b) Alter lens_psfmtf to use the Zernike function. Input coefficient values from Table 8.4. The values are in "waves," so multiply by λ. Generate PSF and MTF results. How do they compare with the Seidel versions?

8.7 Apply the psf_map code to investigate the effects on the PSFs for specific Seidel aberration terms. Keep the exit pupil parameters the same as in the example but look at the following:

(a) 1λ defocus;

(b) 1λ spherical aberration;

(c) 2λ coma;

(d) 1λ astigmatism;

(e) 1λ field curvature; and

(f) $\pm 3\lambda$ distortion.

8.8 Strehl Ratio: The Strehl ratio is another common measure of optical system performance. It is the ratio of the *central* irradiance of the PSF for an aberrated system to the central irradiance of the unaberrated (diffraction-limited) PSF. For an on-axis PSF this can be written as

$$S = \frac{I(0,0)_{ab}}{I(0,0)_{unab}}. \quad (8.18)$$

(a) Start with the lens_psfmtf example code. Set all the aberration coefficients to zero and find the peak value of the PSF. One option is to use max(max(h2)). This value represents I_{unab}.

(b) Find the peak PSF values for each of the following spherical aberration coefficient values: $W_{040} = 0$, 0.1λ, 0.25λ, 0.5λ, 0.75λ, and

1λ. Assume $\lambda = 0.55$ μm and the other aberration coefficients are zero.

(c) Form the Strehl ratios S with the results from parts (a) and (b). Create a plot of S versus the W_{040} coefficient values.

(d) A convenient analytical approximation for the Strehl ratio is[2]

$$S \approx \exp(-k^2 \sigma^2), \qquad (8.19)$$

where k is the wavenumber and σ^2 is the wavefront variance. When only spherical aberration is present, the wavefront variance is given by $\sigma^2 = (1/12 + 1/180)W_{040}$.[2] Plot the analytical and computational Strehl results on the same graph. Are the results consistent?

(e) Compute S for the $f/5$ example lens presented in the `lens_psfmtf` code for the on-axis image point.

8.9 References

1. J. C. Wyant and K. Creath, Basic Wavefront aberration theory for optical metrology, in *Applied Optics and Optical Engineering,* Vol. XI, R. R. Shannon and J. C. Wyant, (eds.), Academic, New York (1992).

2. J. M. Geary, *Introduction to Lens Design with Practical ZEMAX® Examples,* Willmann-Bell, Richmond, VA (2002).

3. E. Dereniak and T. D. Dereniak, *Geometrical and Trigonometric Optics,* Cambridge University Press, Cambridge, U.K. (2008).

4. W. J. Smith, *Modern Optical Engineering,* 4th Ed., McGraw-Hill Professional, New York (2007).

5. J. W. Goodman, *Introduction to Fourier Optics,* 3rd Ed., Roberts & Company, Greenwood Village, CO (2005).

6. J. D. Schmidt, *Numerical Simulation of Optical Wave Propagation with Examples in MATLAB®,* SPIE Press, Bellingham, WA (2010). doi:[10.1117/3.866274].

7. M. C. Roggeman and B. M. Welsh, *Imaging Through Turbulence,* CRC, Boca Raton, FL (1996).

Chapter 9
Partial Coherence Simulation

Except for a brief foray into incoherent imaging in Chapter 7, we have been considering coherent light. Coherence refers to the amount of correlation in the optical field at separate times or separate points within a beam of light. High coherence leads to stationary interference effects such as the "fringing" and "ringing" structures that appeared in the simulated diffraction and imaging results for monochromatic (coherent) light. These kinds of features are absent from the incoherent imaging results.

Coherence is exploited in a variety of optical applications: holography, interferometry, optical coherence tomography, coherent lidar, Fourier transform spectroscopy, and quantum communications, to name a few. We won't get into the details of these topics here, but some applications simply require high coherence (holography) while others also take advantage of the lack of coherence (optical coherence tomography). Sometimes coherence is a noise source—recall the speckled coherent images of Section 7.2.4.

It is convenient to separate coherence into two categories: *temporal* and *spatial*. For illustration, Fig. 9.1 shows a beam of light with some sample points. Temporal coherence refers to the correlation (time average of the products of complex fields) at spatial point P but separated in time by τ. The degree of coherence for a *partial temporal coherent* source changes with τ, typically, decreasing as τ increases. Spatial coherence refers to the correlation of complex fields at the same time but at different transverse points P_1 and P_2. For a *partial spatial coherent* source, the degree of coherence typically decreases with separation distance.

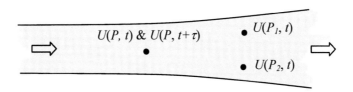

Figure 9.1 Correlation of complex fields at point P separated in time by τ is a measure of temporal coherence. Correlation of fields at points P_1 and P_2 at the same time is a measure of spatial coherence.

Optical coherence theory is a rich statistical discipline, but in this chapter only a few summary results are provided that are specific to the computational discussion. The interested reader is encouraged to explore other resources on this topic.[1-5] As the emphasis here is the propagation of partially coherent light, it is constructive to note that generalized analytic solutions exist for the propagation of nonmonochromatic light. For example, see Section 3.8 in *Introduction to Fourier Optics* by Goodman.[6] Furthermore, there are a number of ways one might go about simulating the propagation of partially coherent radiation (see the introductory remarks by Rydberg and Bengtsson).[7]

The computation approaches presented here are not derived from an analytic solution, but represent a practical strategy based on the concept that partially coherent light is a superposition of irradiance from uncorrelated coherent waves. The simulation of partial temporal coherence and partial spatial coherence is handled in separate ways. However, the approaches have the common theme that a series of propagations or imaging simulations are performed using the coherent methods, and then the irradiance patterns from coherent results are summed to give the partially coherent result.

9.1 Partial Temporal Coherence

9.1.1 Quasi-monochromatic light

Monochromatic light is characterized by a single temporal frequency ν. *Polychromatic* light contains multiple frequencies or a spread in temporal bandwidth $\Delta\nu$. The finite bandwidth corresponds to a loss of temporal coherence. Some sources, such as an incandescent lamp, emit a relatively large range of wavelengths. The focus in this section is on *quasi-monochromatic* light, where $\Delta\nu << \nu_0$. In this case, the bandwidth $\Delta\nu$ is much smaller than the mean, or center frequency ν_0. Lasers are generally quasi-monochromatic sources and their partial temporal coherence characteristics are such that the effects can be noticeable or even exploited in practical applications.

Quasi-monochromatic light can be characterized by a *power spectral density* that describes the relative irradiance contributions of the optical frequencies. This density function is also referred to as the *lineshape*. Common power spectral density functions for quasi-monochromatic light include rectangular, Gaussian, and Lorentzian.[1,2] Here, we consider a normalized Gaussian lineshape given by

$$\hat{S}(\nu) = \frac{1}{\sqrt{\pi}b}\exp\left[-\frac{(\nu-\nu_0)^2}{b^2}\right], \tag{9.1}$$

where b is a width parameter and ν_0 is the center frequency (Hz). The integral of the normalized spectral density over all frequencies is equal to unity. When characterizing the spectral density with a single number it is common to refer to the *linewidth* $\Delta\nu$, which is a full width at half-maximum (FWHM) measure (Hz) of the spectral density function. Considering the half-maximum value for Eq. (9.1), the following can be derived:

$$b = \frac{\Delta v}{2\sqrt{\ln 2}}. \tag{9.2}$$

A measure of the temporal coherence of the optical field is the *complex degree of temporal coherence* $\gamma(\tau)$. It is the normalized correlation of the field where τ is the time delay between correlation samples. The normalized spectral density and complex degree of temporal coherence are linked by a Fourier transform relationship:[1]

$$\gamma(\tau) = \int_0^\infty \hat{S}(v) \exp(-j2\pi v\tau) dv, \tag{9.3}$$

where $\hat{S}(v)$ is defined for positive frequencies. Performing the transform on the spectral density in Eq. (9.1) gives

$$\gamma(\tau) = \exp\left[-\left(\frac{\pi \Delta v \tau}{2\sqrt{\ln 2}}\right)^2\right] \exp(-j2\pi v_0 \tau). \tag{9.4}$$

Our concern is with $|\gamma(\tau)|$, where it can be shown that

$$0 \le |\gamma(\tau)| \le 1. \tag{9.5}$$

Perfect coherence is indicated by $|\gamma(\tau)| = 1$ and incoherence is indicated by $|\gamma(\tau)| = 0$. Figure 9.2 illustrates that as the absolute value of τ increases, $|\gamma(\tau)|$ decreases, so further time separation means less coherence.

A single-value characterization of temporal coherence is the *coherence time* τ_c. For a Gaussian lineshape the coherence time is defined as[1]

$$\tau_c = \frac{0.664}{\Delta v}. \tag{9.6}$$

For a fixed spatial position in the path of a beam of light, the field is highly correlated over a time that is much less than the coherence time. For a time that is on the order of the coherence time, the correlation is significantly reduced but may still cause some noticeable interference effects. The distance the beam travels during the coherence time is known as the *coherence length*

$$\ell_c = c\tau_c. \tag{9.7}$$

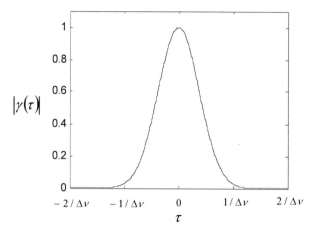

Figure 9.2 Magnitude of the complex degree of coherence corresponding to Eq. (9.4). The sign of τ depends on which sample point is delayed.

For example, suppose $\lambda_0 = 650$ nm and $\Delta v = 2$ GHz (a wavelength and linewidth that are typical of a laser diode). In this case, $\tau_c = 3.32 \times 10^{-10}$ s and $\ell_c = 10$ cm. Imagine splitting this beam of light equally and "delaying" one part by making it travel an extra distance of 10 cm. If the two beams are recombined, the contrast in the resulting interference fringes will be low.

9.1.2 Partial temporal coherence simulation approach

For simulation purposes the total irradiance in an x–y plane can be modeled as[8]

$$I(x,y) = \int_{-\infty}^{\infty} \hat{S}(v) I(x,y;v) dv, \qquad (9.8)$$

where $I(x,y;v)$ is a spectral irradiance (W/m²·Hz). On the computer the spectral content can be modeled as a sum of discrete components:

$$I(x,y) \approx \sum_{n=1}^{N} \hat{S}(v_n) I(x,y;v_n) \delta v, \qquad (9.9)$$

where n indexes the components, N is the number of components, and δv is the frequency interval between components. Figure 9.3 shows an example of a normalized, sampled, power spectral density function. In Eq. (9.8) a simple continuous power spectral density function is assumed. But with the appropriate model more complicated density functions can be handled, such as a series of distinct mode features that are found in some laser sources.

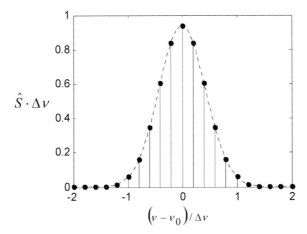

Figure 9.3 Normalized optical power spectral density with sampled components.

Thus, the simulation approach is to propagate (or image with) a series of fields at different frequencies, form the irradiance for each, weigh each by the power spectral density, and sum the patterns to get the partially coherent result.

This approach ignores frequency component cross-correlations, in other words, interference between frequency components. Cross-correlation could result in temporal beating effects in the irradiance rather than a stationary pattern. The beat frequencies would be equal to the differences of the component frequencies but may not be detectable by the sensor depending on the sensor's temporal response. The correlations may also be weak. Although not fully justified, this is a common assumption in practice that seems to give credible results.[7,9] See the related discussion in Section 9.3.

To observe partial temporal coherence effects, some type of differential delay must occur between different segments of the field in traveling from the source to the observation plane. The delay can come about in a variety of natural ways; for example, multipath scattering by something in the propagation medium or reflection off of surfaces with a depth profile. A delay can also be arranged in an optical setup.

9.1.3 Partial temporal coherence example

Consider the arrangement in Fig. 9.4. Two parallel beams, both circular in shape with radius w, enter from the left with a center-to-center separation of Δs. They are assumed to have come from the same source with no temporal delay between them. Using mirrors, the lower leg takes a detour that adds a distance of Δd to its path relative to the top beam. This can be interpreted as a relative time delay of $\tau = \Delta d/c$. The beams are focused by a lens of focal length f to form a Fraunhofer pattern. This is essentially Young's double-slit arrangement, except with holes and a lens. Diffractive effects to the left of the lens are ignored, so circle functions are assumed for the beams at the lens.

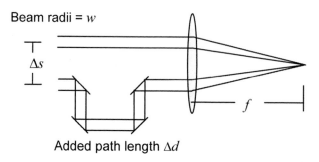

Figure 9.4 Two-beam temporal coherence arrangement.

Choose some parameters for the simulation; for example, $\lambda_0 = 650$ nm, $\Delta v = 2$ GHz, $w = 1$ mm, $\Delta s = 5$ mm, $f = 0.25$ m, and $\Delta d = 5$ cm. Here is the code for pc_temp:

```
1   % pc_temp partial temporal coherence example
2
3   lambda0=650e-9;         %center wavelength (m)
4   c=3e8;                  %speed of light
5   k0=2*pi/lambda0;        %center wavenumber
6   nu0=c/lambda0;          %center frequency
7
8   % Gaussian lineshape parameters
9   N=51;                   %number of components (odd)
10  delnu=2e9;              %spectral density FWHM (Hz)
11  b=delnu/(2*sqrt(log(2))); %FWHM scaling
12  dnu=4*delnu/N;          %freq interval
13
14  % source plane parameters
15  L1=50e-3;               %source plane side length
16  M=250;                  %# samples (even)
17  dx1=L1/M;               %sample interval
18  x1=-L1/2:dx1:L1/2-dx1;  %source coords
19  x1=fftshift(x1);        %shift x coord
20  [X1,Y1]=meshgrid(x1,x1);
21
22  % beam parameters
23  w=1e-3;                 %radius
24  dels=5e-3;              %transverse separation
25  deld=5e-2;              %delay distance
26  f=0.25;                 %focal dist for Fraunhofer
27  lf=lambda0*f;
28
29  % loop through lines
30  I2=zeros(M);
31  for n=1:N
32      % spectral density function
33      nu=(n-(N+1)/2)*dnu+nu0;
```

```
34      S=1/(sqrt(pi)*b)*exp(-(nu-nu0)^2/b^2);
35      k=2*pi*nu/c;
36      % source
37      u=circ(sqrt((X1-dels/2).^2+Y1.^2)/w)...
38          +circ(sqrt((X1+dels/2).^2+Y1.^2)/w)...
39          *exp(j*k*deld);
40      % Fraunhofer pattern
41      u2=1/lf*(fft2(u))*dx1^2;
42      % weighted irradiance and sum
43      I2=I2+S*(abs(u2).^2)*dnu;
44  end
45
46  I2=ifftshift(I2);   %normalize/center irradiance
47  x2=(-1/(2*dx1):1/L1:1/(2*dx1)-1/L1)*lf;  %obs coords
48  y2=x2;
49
50  figure(1)           %irradiance image
51  imagesc(x2,y2,I2);
52  xlabel('x (m)'); ylabel('y (m)');
53  axis square; axis xy; colormap('gray');
54
55  figure(2)           %irradiance profile
56  plot(x2,I2(M/2+1,:));
57  xlabel('x (m)'); ylabel('Irradiance');
```

Some comments about this code are as follows:

(a) Line 8: The modeling accuracy for the spectral density function depends on the number of discrete spectral lines N and the frequency interval dnu ($\delta \nu$). After some experimentation N = 51 and dnu = 4*delnu/N were selected. The 51 lines span $4\Delta \nu$, or four times the FWHM of the spectral density. This is the span displayed in Fig. 9.2. These values give results that are consistent with theory in this case. Of course, the more lines and greater span, the better the discrete spectral model follows the analytic model, but execution time can become a problem.

(b) Line 15: The source plane side length L1 = 50 mm is selected to sample the circ beams adequately and give a reasonable pattern size in the observation plane. No need to worry about the Fraunhofer phase sampling and aliasing.

(c) Line 19: fftshift is applied to x1 so the field u is directly calculated in the shifted position. This avoids the use of fftshift within the for loop (a time saver).

(d) Line 31: The `for` loop is used to compute and sum the N Fraunhofer irradiance patterns, each weighted by the discrete spectral density function S.

(e) Line 39: The two holes are oriented along the x axis. The delay distance Δd is set by the variable `deld`. For `u`, the field at the input to the lens, the effect of the delay is modeled by the complex exponential term that includes the optical path difference, $\exp(jk\Delta d)$.

(f) Line 41: The front-end Fraunhofer phase terms are ignored. The center wavelength `lambda0` is used for scaling the Fraunhofer pattern in both the multiplicative `1/lf` term and the Fraunhofer coordinates `x2` (line 47). The slight spread in wavelength for the quasi-monochromatic light causes insignificant effects in these terms (unlike the optical path difference term).

(g) Line 43: The integration approximation for Eq. (9.9) is performed.

Results for the code are shown in Fig. 9.5. With perfectly coherent beams the fringe "nulls" will reach zero, but clearly the interference is not complete. This is expected for a coherence length of $\ell_c = 10$ cm and path difference of $\Delta d = 5$ cm. For this particular case, $\tau_c = \Delta d/c = 1.67 \times 10^{-10}$ s and the modulus of the complex degree of temporal coherence is calculated to be

$$|\gamma(\tau)| = \exp\left[-\left(\frac{\pi \Delta \nu \tau}{2\sqrt{\ln 2}}\right)^2\right] = 0.6733.$$

An analytic solution for the irradiance pattern in this case is given by[1]

$$I_2(x_2, y_2) = \frac{2}{(\lambda_0 f)^2}\left[w^2 \frac{J_1\left(2\pi \frac{w}{\lambda_0 f}\sqrt{x_2^2 + y_2^2}\right)}{\frac{w}{\lambda_0 f}\sqrt{x_2^2 + y_2^2}}\right]^2$$

$$\times \left[1 + \left|\gamma\left(\frac{\Delta d}{c}\right)\right|\cos\left(2\pi \frac{\Delta s}{\lambda_0 f}x_2 + k_0 \Delta d\right)\right], \quad (9.10)$$

and from Eq. (9.4)

$$\left|\gamma\left(\frac{\Delta d}{c}\right)\right| = \exp\left[-\left(\frac{\pi \Delta \nu \Delta d/c}{2\sqrt{\ln 2}}\right)^2\right]. \quad (9.11)$$

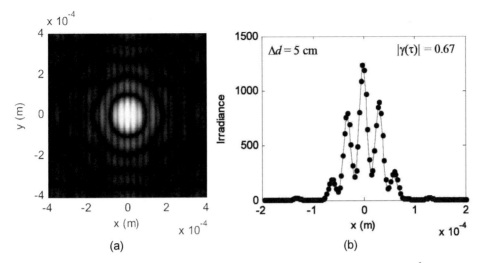

Figure 9.5 Partial temporal coherence results for pb_temp, where $\Delta \nu = 2$ GHz ($\ell_c = 10$ cm) and $\Delta d = 5$ cm: (a) irradiance image and (b) x axis profile. The x-axis in (b) is expanded for a better view of the center of the pattern. Points in (b) are from the analytic result.

The points displayed in Fig. 9.5(b) are generated using Eq. (9.10). They show a close fit to the simulation result.

The panels in Fig. 9.6, created with pc_temp, show the transition from high to low contrast fringes in a series of Fraunhofer images and profiles for Δd ranging from 0 to 50 cm. Essentially all coherent interference between the beams is lost for $\Delta d = 50$ cm and the Airy pattern shape for a single beam appears—although the relative irradiance is doubled because of the two beams. The "lumpiness" in the Airy pattern rings, most notable in the $\Delta d = 50$ cm case, is due to the discrete edge sampling of the original circle functions.

9.2 Partial Spatial Coherence

9.2.1 Stochastic transmittance screen

The field from an ideal point source is perfectly spatially coherent. If the field from the source is observed at two points in space, the amplitudes will be perfectly correlated. But with a spatially extended collection of independently radiating point sources involving different frequencies and amplitudes, the correlation between the field at the two observation points decreases. Similarly, if the field is somehow affected randomly in time along the different propagation paths, then spatial coherence will decrease. Conversely, as light travels a long distance from a source, the wavefront becomes more planar and spatial coherence increases. For example, light from distant stars has high spatial coherence.

A measurement of spatial coherence is the *complex coherence factor* μ_{12}. It is a normalized correlation of the field at two points, 1 and 2. We are concerned with $|\mu_{12}|$, where it can be shown that

$$0 \le |\mu_{12}| \le 1. \tag{9.12}$$

The correlation between the field samples is performed over time. Suppose the two points are separated in the x–y plane by the values x' and y', respectively (Fig. 9.7). For simulation purposes we consider a simple model with a deterministic part of the field $U_0(x, y)$ and a stochastic temporal component introduced through a complex transmittance screen $t_A(x, y; t)$. Thus, the field is modeled by

$$U(x, y; t) = U_0(x, y) t_A(x, y; t). \tag{9.13}$$

A quantity of interest is the time-averaged spatial autocorrelation function of the transmittance screen, given as

$$R(x', y') = \langle t_A(x + x', y + y'; t) t_A^*(x, y; t) \rangle, \tag{9.14}$$

where the angle brackets indicate the average and * is the complex conjugate.

The complex coherence factor is equivalent to the normalized screen autocorrelation function[1]

$$\mu_{12} = \frac{R(x', y')}{R(0,0)}. \tag{9.15}$$

A commonly used form for the spatial correlation function is a Gaussian

$$R(x', y') = \exp\left(-\frac{x'^2 + y'^2}{\ell_{cr}^2}\right), \tag{9.16}$$

where ℓ_{cr} is a measure of the spatial or *transverse coherence length*.

9.2.2 Partial spatial coherence simulation approach

The simulation approach is outlined as follows: a spatially random transmittance screen is applied to a deterministic beam field, the result is propagated, and the irradiance is formed. The process is repeated many times with different realizations of the screen and the resulting irradiance patterns are averaged to produce the partial spatial coherent result. The procedure is illustrated in Fig. 9.8.

A convenient form of a transmittance function is a complex phasor given by

$$t_A(x, y) = \exp[j\phi(x, y)], \tag{9.17}$$

where $\phi(x,y)$ is a spatially correlated, random phase function that is often called a *phase screen*. The transmittance function of Eq. (9.17) preserves the

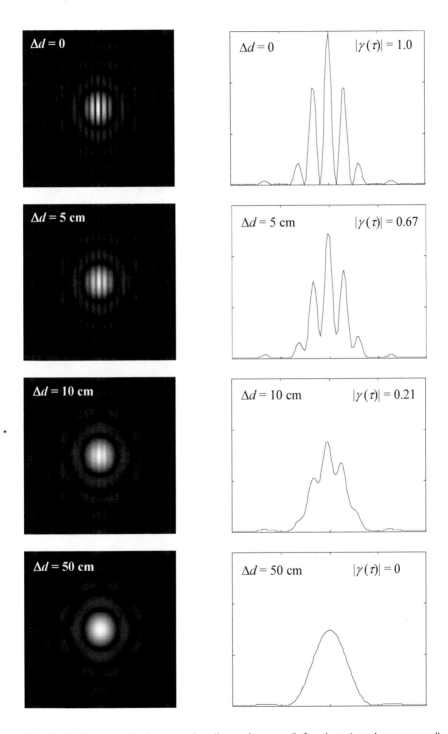

Figure 9.6 Partial temporal coherence irradiance images (left column) and corresponding x-axis profiles (right column) for $\Delta\nu = 2$ GHz ($\ell_c = 10$ cm) and $\Delta d = 0$–50 cm. Axis scaling is the same as for Fig. 9.5. Airy ring "lumpiness" is due to the initial circle-beam sampling.

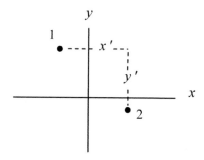

Figure 9.7 Illustration of field sample positions for spatial coherence correlation.

magnitude of the deterministic part of the field. The phase is modeled by

$$\phi(x, y) = r(x, y) \otimes f(x, y), \qquad (9.18)$$

where $r(x,y)$ is a spatially uncorrelated, or "delta correlated" random signal with variance σ_r^2, and $f(x,y)$ is a Gaussian response function that acts to "smooth" the random signal and create spatial correlation in the phase signal. If the Gaussian function is given by

$$f(x, y) = \frac{1}{\pi \sigma_f^2} \exp\left(-\frac{x^2 + y^2}{\sigma_f^2}\right), \qquad (9.19)$$

where σ_f is the width parameter, then the autocorrelation of Eq. (9.17) can be shown to be[10]

$$R(x', y') = \exp\left\{-\frac{\sigma_r^2}{2\pi \sigma_f^2}\left[1 - \exp\left(-\frac{x'^2 + y'^2}{2\sigma_f^2}\right)\right]\right\}. \qquad (9.20)$$

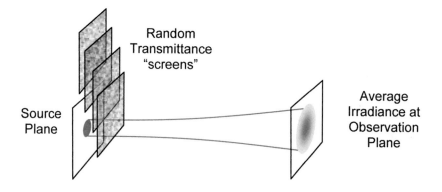

Figure 9.8 Illustration of partial spatial coherence simulation approach.

Furthermore, if

$$\frac{\sigma_r^2}{2\pi\sigma_f^2} \gg 1, \qquad (9.21)$$

a Gaussian correlation function is obtained

$$R(x', y') \approx \exp\left(-\frac{x'^2 + y'^2}{\ell_{cr}^2}\right), \qquad (9.22)$$

where the transverse coherence length is given by

$$\ell_{cr}^2 = \frac{4\pi\sigma_f^4}{\sigma_r^2}. \qquad (9.23)$$

To create $t_A(x,y)$ in the computer, $\phi(x,y)$ is first constructed based on Eq. (9.18). This can be done using a Fourier filtering method in the frequency domain. If $|\Phi|^2$ is the power spectrum of ϕ, σ_r^2 is the variance of the random signal r and $F = \Im\{f\}$, then using the property of a linear filter with a random input signal we write[11]

$$|\Phi(f_X, f_Y)|^2 = \sigma_r^2 |F(f_X, f_Y)|^2. \qquad (9.24)$$

The operation performed in the frequency domain needs to satisfy Eq. (9.24). In the computer with discrete arrays and indices p and q, we do the following:[10,12]

$$\Phi(p\Delta f_X, q\Delta f_y) = \tilde{r}(p,q)\frac{\sigma_r}{(\Delta f_X \Delta f_Y)^{1/2}} F(p\Delta f_X, q\Delta f_Y), \qquad (9.25)$$

where $\tilde{r}(p,q)$ is a zero-mean, unit variance random array. The term $(\Delta f_X\Delta f_Y)^{1/2}$ arises because a sample interval "area" associated with the variance of the samples needs to be normalized. For uniform sampling $(\Delta f_X\Delta f_Y)^{1/2} = \Delta f_X$. For the Gaussian correlation case, the transform of Eq. (9.19) gives the following filter:

$$F(f_X, f_Y) = \exp\left[-\pi^2\sigma_f^2\left(f_X^2 + f_Y^2\right)\right]. \qquad (9.26)$$

So, the computational approach to get a random realization of ϕ is to fill an array with random values $\tilde{r}(p,q)$, multiply by $\sigma_r/\Delta f_X$, multiply by the filter response $F(p\Delta f_X, q\Delta f_Y)$, and take the inverse FFT.

One final trick is to make $\tilde{r}(p,q)$ a random complex array, where the real and imaginary parts are independent Gaussian random values. Thus, Φ generated via Eq. (9.25) is complex. Taking the inverse FFT we get ϕ, where the real and imaginary parts are independent random arrays, each with the correlation specified by Eq. (9.22)—so, two random phase screens for the price of one FFT! Finally, the screens are inserted in complex phasors to make two discrete realizations of the transmittance function t_A.

9.2.3 Partial spatial coherence example

To illustrate partial spatial coherence effects, the same optical arrangement as for the temporal case is used, but the path delay is removed (Fig. 9.9). In this case the limited spatial coherence does the job of reducing the fringe contrast in the Fraunhofer pattern. The same parameters as in the temporal case are used for the simulation: $\lambda = 650$ nm, $w = 1$ mm, $\Delta s = 5$ mm, and $f = 0.25$ m. The transverse coherence length is selected as $\ell_{cr} = 8$ mm, which is on the order of Δs and, therefore, should produce an obvious reduction in coherence between the two beams. In this case,

$$|\mu_{12}| = \exp(-\Delta s^2 / \ell_{cr}^2) = 0.677.$$

The code for pc_spatial is as follows:

```
1   % pc_spatial partial spatial coherence example
2
3   lambda=650e-9;      %center wavelength (m)
4
5   L1=50e-3;           %source plane side length
6   M=250;              %# samples (even)
7   dx1=L1/M;           %sample interval
8   x1=-L1/2:dx1:L1/2-dx1; %source coords
9   x1=fftshift(x1);    %shift x coord
10  [X1,Y1]=meshgrid(x1,x1);
11
12  % beam parameters
```

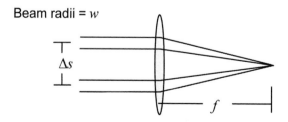

Figure 9.9 Two-beam spatial coherence arrangement.

```
13  w=1e-3;                %radius
14  dels=5e-3;             %transverse separation
15  f=0.25;                %Fraunhofer focal distance
16  lf=lambda*f;
17
18  % partial spatial coherence screen parameters
19  N=100;                 %number of screens (even)
20  Lcr=8e-3;              %spatial correlation length
21  sigma_f=2.5*Lcr;       %Gaussian filter parameter
22  sigma_r=sqrt(4*pi*sigma_f^4/Lcr^2); %random std
23
24  dfx1=1/L1;
25  fx1=-1/(2*dx1):dfx1:1/(2*dx1)-dfx1;
26  fx1=fftshift(fx1);
27  [FX1,FY1]=meshgrid(fx1,fx1);
28
29  % source field
30  u1=circ(sqrt((X1-dels/2).^2+Y1.^2)/w)...
31      +circ(sqrt((X1+dels/2).^2+Y1.^2)/w);
32  % filter spectrum
33  F=exp(-pi^2*sigma_f^2*(FX1.^2+FY1.^2));
34
35  % loop through screens
36  I2=zeros(M);
37  for n=1:N/2
38      % make 2 random screens
39      fie=(ifft2(F.*(randn(M)+j*randn(M)))...
40          *sigma_r/dfx1)*M^2*dfx1^2;
41      % Fraunhofer pattern applying screen 1
42      u2=1/lf*(fft2(u1.*exp(j*real(fie))))*dx1^2;
43      I2=I2+abs(u2).^2;
44      % Fraunhofer pattern applying screen 2
45      u2=1/lf*(fft2(u1.*exp(j*imag(fie))))*dx1^2;
46      I2=I2+abs(u2).^2;
47  end
48
49  I2=ifftshift(I2)/N;    %normalize & center irradiance
50  x2=(-1/(2*dx1):1/L1:1/(2*dx1)-1/L1)*lf; %obs coords
51  y2=x2;
52
53  figure(1)              %irradiance image
54  imagesc(x2,y2,I2);
55  xlabel('x (m)'); ylabel('y (m)');
56  axis square; axis xy;
57  colormap('gray');
58
59  figure(2)              %irradiance slice
60  plot(x2,I2(M/2+1,:));
61  xlabel('x (m)'); ylabel('Irradiance');
```

Comments for this code are as follows:

(a) Line 9: `fftshift` is applied to `x1` so the field `u` is calculated in the shifted position, which avoids the use of `fftshift` within the `for` loop.

(b) Line 19: The number of phase screen realizations is the same kind of parameter as the number of spectral lines in the partial temporal coherence simulation. The simulation result tends to converge to the analytic result as a function of N, but execution run times get long with large N. In this case N=100 was chosen experimentally. This value gave results that appeared to be "smooth" and followed the analytic result.

(c) Line 26: Same comment as for line 9 for `fx1`.

(d) Line 37: Loop through the screens.

(e) Line 39: The random number generator `randn` creates an array of zero-mean normally distributed random values. Scaling the phase function $\phi(x,y)$ appropriately requires several factors: `/dfx1` comes from Eq. (9.25); `M^2` counteracts the $1/MN$ factor that accompanies the `ifft2` function; `dfx1^2` is applied to correctly approximate the inverse Fourier transform integral.

Regarding lines 21–23, there is some flexibility in the choice of σ_r and σ_f for a given value of ℓ_{cr}. The approach used here for selecting σ_r and σ_f is to first invert Eq. (9.21) to get $2\pi\sigma_f^2/\sigma_r^2 \ll 1$. This result is combined with Eq. (9.23) to give $\sigma_f^2 \gg \ell_{cr}^2/2$ and, finally, the choice of $\sigma_f = 2.5\,\ell_{cr}$ is made. This approach selects roughly the smallest value of σ_f that allows Eq. (9.21) to be satisfied. For this example, $\ell_{cr} = 8$ mm, thus $\sigma_f = 20$ mm, and from Eq. (9.23), $\sigma_r = 177.2$ mm. When σ_f is too large, say roughly $\sigma_f > L/2$, then the passband of the filter F is only defined by a few samples in the frequency domain, and the screen generated can become inaccurate. Large correlation lengths ℓ_{cr} lead to this problem, and the failure mode of the simulation in this situation is to output the perfectly coherent result. Sometimes choosing σ_f to be smaller, for example, $\sigma_f = 2\,\ell_{cr}$ or less, can help with this problem. This approach can still yield good results even though Eq. (9.21) is violated.

Before discussing the irradiance results for the `pc_spatial` script, it is informative to look at a few of the random phase screens that are applied to the source plane. Two realizations generated with the `pc_spatial` parameters are shown in Fig. 9.10. The phase values and apparent variations seem reasonable for $\sigma_f = 20$ mm. One hundred realizations of the phase screens are applied in `pc_spatial`. Be aware that if phase excursions for the screens become large,

Partial Coherence Simulation 185

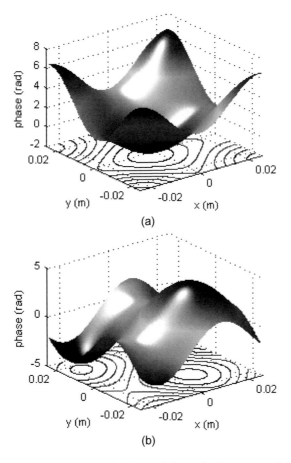

Figure 9.10 Two random realizations of $\phi(x, y)$ for the pc_spatial example.

the phase can become undersampled, and aliasing can result, as discussed for aberrated wavefronts in Section 8.6.

Executing `pc_spatial` generates the irradiance results in Fig. 9.11. The partial spatial coherence creates a reduction in the interference fringe contrast that is similar to the temporal coherent case (Fig. 9.5). The analytic solution for this case is given by

$$I_2(x_2, y_2) \approx \frac{2}{(\lambda f)^2} \left[w^2 \frac{J_1\left(2\pi \frac{w}{\lambda f} \sqrt{x_2^2 + y_2^2}\right)}{\frac{w}{\lambda f} \sqrt{x_2^2 + y_2^2}} \right]^2 \qquad (9.27)$$

$$\times \left[1 + |\mu_{12}| \cos\left(2\pi \frac{\Delta s}{\lambda f} x_2\right)\right].$$

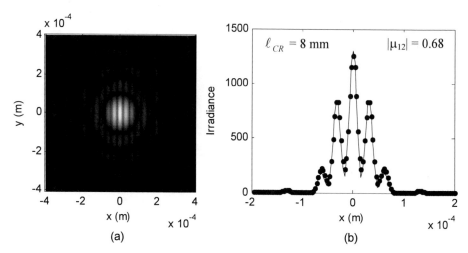

Figure 9.11 Partial spatial coherence results for pc_spatial where ℓ_{cr} = 8 mm: (a) irradiance image and (b) x-axis profile. The x axis in (b) is expanded for a better view of the center of the pattern. Points in (b) are from the analytic result.

The result in Eq. (9.27) breaks down if the transverse coherence length is on the order of, or smaller than, the beam diameters.

9.3 Reducibility, Number of Spectral Components, and Phase Screens

Temporal and spatial coherence are treated separately in the developments of the previous sections. However, the general form of the complex degree of coherence $\gamma_{12}(\tau)$ encompasses both temporal and spatial field correlations and can include cross-correlations of the frequency components. Coherence separation implies $\gamma_{12}(\tau)$ can be factored into a temporal term and spatial term product, or[1]

$$\gamma_{12}(\tau) = \mu_{12}\gamma(\tau). \qquad (9.28)$$

This factoring is possible if the light is *cross-spectrally pure*. Roughly speaking, cross-spectral purity means the coherence properties of the light are position independent. In practice, this can be a reasonable assumption for many optical sources, for example, lasers. However, exceptions exist such as light from a source with an angle-dependent spectral density.[1] A coherence function that can be factored, as in Eq. (9.28), is said to be *reducible*. All examples illustrated in this chapter assume the source coherence function is reducible.

Little discussion was provided with regard to the number of spectral components in the temporal coherence modeling. The number was simply chosen through experimentation. This and other related issues have been considered in more detail by other authors. For example, Rydberg and Bengtsson discuss spectral sampling and extend the method to consider pulsed optical sources.[7]

Similarly, the parameter values and the number of applied phase screens for spatial coherence simulation were chosen through experimentation, and a criterion for the minimum number of screens was not proposed. Spatial coherence modeling is a more recent activity, and some of these issues have yet to be fully investigated. This leaves plenty of room for more study and development.

9.4 Exercises

9.1 A rectangular lineshape is given by

$$\hat{S}(\nu) = \frac{1}{\Delta \nu} \text{rect}\left(\frac{\nu - \nu_0}{\Delta \nu}\right), \tag{9.29}$$

and the coherence time for this lineshape is[1]

$$\tau_c = \frac{1}{\Delta \nu}. \tag{9.30}$$

(a) Alter the `pc_temp` code to use this lineshape. Keep the simulation parameters the same ($\lambda_0 = 650$ nm, $\Delta \nu = 2$ GHz, $w = 1$ mm, $\Delta s = 5$ mm, $f = 0.25$ m, and $\Delta d = 5$ cm). Generate a simulated irradiance pattern and profile.

(b) Derive an expression for $|\gamma(\tau)|$. Find the value for $|\gamma(\tau)|$ given the parameters in step (a).

(c) Compare the simulation result in step (a) with the analytic result.

9.2 Fringe shift: A sensitive method for determining the path length difference between two *coherent* beams is to examine the apparent "translation" of fringes created in the Fraunhofer pattern. Start with the `pc_temp` example.

(a) Alter the script to generate the Fraunhofer pattern for a coherent beam ($\lambda_0 = 650$ nm). Essentially remove the looping.

(b) Create irradiance profiles for $\Delta d = 0$, $\lambda_0/4$, $\lambda_0/2$, and $\lambda_0 3/4$. Is there a limitation in terms of the amount of optical path difference that can be measured using this technique?

9.3 If the two beams in Fig. 9.4 have different field magnitudes given by A_1 and A_2, a quantity known as the *visibility* is defined as

$$V = \frac{2 A_1 A_2}{A_1^2 + A_2^2} \gamma\left(\frac{\Delta d}{c}\right). \tag{9.31}$$

Consider the `pc_temp` example ($\Delta d = 5$ cm) and assume $A_1 = 0.75$, $A_2 = 0.20$.

(a) What is the value of the visibility?

(b) Alter the `pc_temp` script and generate an irradiance profile.

(c) With the help of Exercise 4.5 and combining Eqs. (9.31) and (9.10), "propose" an analytic expression for the Fraunhofer pattern and test it against the simulation result.

9.4 Implement the `pc_spatial` simulation and the analytic solution [Eq. (9.27)] for irradiance profile comparison. Generate irradiance results for $\ell_{cr} = 15$, 8, 4, and 1 mm. What are the corresponding values of $|\mu_{12}|$? Recall that if $\sigma_f > L/2$ the simulation can become inaccurate. For the 15-mm case, try adjusting the 2.5 multiplier in line 21 to a smaller value. More screen realizations may be needed to get a good 1-mm result. Can you explain the 1-mm result?

9.5 Replace the two beams in the `pc_spatial` example with a single 5 mm × 2 mm rectangular aperture (illuminated with a unit amplitude plane wave). Examine the Fraunhofer patterns for ℓ_{cr} ranging from 20 mm to 1 mm. Can you explain the result?

9.6 Gaussian Schell-model beam: A Gaussian Schell-model (GSM) beam has a Gaussian amplitude envelope and a Gaussian spatial correlation function. This beam has been studied extensively for laser communication applications.[13] The irradiance of the GSM beam is given by[2,10]

$$I_2(x, y) = \frac{I_0}{\Delta^2} \exp\left[-\frac{2(x^2 + y^2)}{w_0^2 \Delta^2}\right], \tag{9.32}$$

where

$$\Delta = \left[1 + \left(\frac{2z}{kw_0^2}\right)^2 \left(1 + 2\frac{w_0^2}{\ell_{cr}^2}\right)\right]^{1/2}, \tag{9.33}$$

and w_0 is the initial beam size parameter. The source beam field is given by

$$U_1 = \exp\left(-\frac{x^2 + y^2}{w_0^2}\right). \tag{9.34}$$

Develop a propagation simulation for a GSM beam. The parameters are $w_0 = 2$ cm, $\lambda = 1.06$ μm, $z = 1$ km, and $\ell_{cr} = 1$ cm. Use an array size of $M = 256$. Choose other necessary parameters. Experiment with the number of random screen realizations. The more, the better, in terms of result accuracy—but execution time can get long.

(a) Plot irradiance profiles of the initial beam at the source plane and the perfectly coherent beam and the GSM beam at the observation plane.

(b) Compare the GSM beam profile with the analytic theory.

9.5 References

1. J. W. Goodman, *Statistical Optics*, Wiley-Interscience, New York (1985).
2. L. Mandel and E. Wolf, *Optical Coherence and Quantum Optics*, Cambridge University Press, Cambridge, U.K. (1995).
3. B. E. A. Saleh and M. C. Teich, *Fundamentals of Photonics*, 2nd Ed., Wiley-Interscience, New York (2007).
4. E. Hecht, *Optics*, 4th Ed., Addison-Wesley, Reading, MA (2002).
5. K. K. Sharma, *Optics: Principles and Applications*, Academic, London (2006).
6. J. W. Goodman, *Introduction to Fourier Optics*, 3rd Ed., Roberts & Company, Greenwood Village, CO (2005).
7. C. Rydberg and J. Bengtsson, "Efficient numerical representation of the optical field for the propagation of partially coherent radiation with a specified spatial and temporal coherence function," *J. Opt. Soc. Am. A,* **23,** 1616–1625 (2006).
8. D. G. Voelz, K. A. Bush, and P. S. Idell, "Illumination coherence effects in laser-speckle imaging: Modeling and experimental demonstration," *Appl. Opt.,* **36,** 1781–1788 (1997).
9. P. Polynkin, A. Peleg, L. Klein, T. Rhoadarmer, and J. Moloney, "Optimized multiemitter beams for free-space optical communications through turbulent atmosphere," *Opt. Lett.,* **32,** 885–887 (2007).
10. X. Xiao and D. Voelz, "Wave optics simulation approach for partial spatially coherent beams," *Opt. Express,* **14,** 6986–6992 (2006).
11. A. Papoulis and S. Unnikrishna Pillai, *Probability, Random Variables, and Stochastic Processes*, 4th Ed., McGraw-Hill, New York (2002).
12. H. A. Makse, S. Havlin, M. Schwartz, and H. E. Stanley, "Method for generating long-range correlations for large systems," *Phys. Rev. E,* **53,** 5445–5449 (1996).
13. J. C. Ricklin and F. M. Davidson, "Atmospheric optical communication with a Gaussian Schell beam," *J. Opt. Soc. Am. A,* **20,** 856–866 (2003).

Appendix A
Fresnel Propagator Chirp Sampling

The transfer function (TF) H and impulse response (IR) h Fresnel propagators are introduced in Chapter 5. Although analytically identical, the methods produce different results with sampled data. The culprit is the sampled chirp functions that are a part of H and h. In this appendix, sampling regimes are derived for the chirp functions, and it is shown that the sense of the regimes is reversed for the two methods. Both methods involve FFT and FFT^{-1} operations, so the effects of performing these discrete transforms on the sampled H and h are illustrated.

A.1 Fresnel Transfer Function Sampling

The Fresnel transfer function

$$H(f_X, f_Y) = e^{jkz} \exp\left[-j\pi\lambda z\left(f_X^2 + f_Y^2\right)\right], \tag{A.1}$$

contains a complex exponential term with a phase function whose absolute value increases with the square of the frequency variables. This type of function is referred to as a "chirp" function, a label that was originally applied to similar functions in the temporal or spatial domains. Sampling a chirp function, as required for a propagation simulation, can be problematic because of the increasing slope of the phase with frequency.[1] Only the phase of the chirp term in Eq. (A.1) is a function of frequency, so extracting the phase gives

$$\phi_H(f_X, f_Y) = -\pi\lambda z\left(f_X^2 + f_Y^2\right). \tag{A.2}$$

Only one transverse direction needs to be examined, as the sampling constraints for the two orthogonal variables can be evaluated separately. For a uniform sample interval of Δf_X the criterion for an unambiguous representation of the phase when it is encoded in a modulo-2π format, which is the case for a complex exponential term, can be written as

$$\Delta f_X \left| \frac{\partial \phi_H}{\partial f_X} \right|_{max} \leq \pi. \qquad (A.3)$$

This expression states that the maximum change in the absolute phase must be no more than π between any two adjacent samples (see, for example, reference 2). If this constraint is violated, then aliased phase values result. The slope is found to be $\partial \phi_H / \partial f_X = -2\pi \lambda z f_X$, and since λ and z are constants for a given propagation, the maximum slope occurs when f_X is a maximum ($f_{X\,max}$). Inserting this information into Eq. (A.3) and solving for Δf_X gives the following criterion for the sample interval:

$$\Delta f_X \leq \frac{1}{\lambda z 2 |f_{X max}|}. \qquad (A.4)$$

Although sampling of the transfer function occurs in the frequency domain, it is helpful to consider the corresponding sampling in the spatial domain. Assuming the frequency and spatial domain sampling are related through the scaling properties of the FFT, then $\Delta f_X = 1/L$ and $|f_{X max}| = 1/2\Delta x$, where L is the side length and Δx is the sample interval in the spatial domain. Substituting these relations in Eq. (A.4) and solving for Δx yields the following spatial domain criterion:

$$\Delta x \geq \frac{\lambda z}{L}. \qquad (A.5)$$

With reference to Eq. (A.5), three sampling regimes for the transfer function H are now discussed.

A.1.1 Oversampled transfer function

The transfer function is *oversampled* when

$$\Delta x > \frac{\lambda z}{L}. \qquad (A.6)$$

This is the case for a relatively "short" propagation distance or wavelength. Figure A.1(a) shows an example profile of the argument of H when H is oversampled. The argument of H is unwrapped to allow comparison with the analytic phase ϕ_H of Eq. (A.2). The two curves in Fig. A.1(a) are identical, thus the condition in Eq. (A.6) usually provides acceptable simulation results. However, looking further at this case, consider that U_2 can be described by

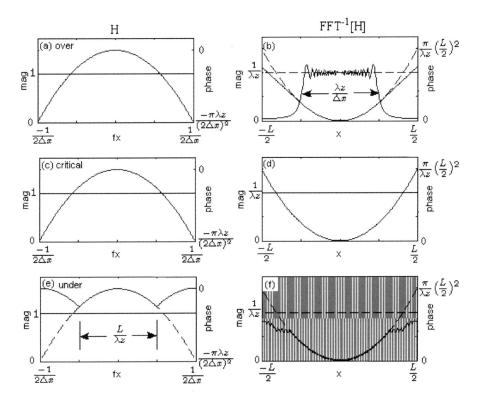

Figure A.1 Magnitude and phase (unwrapped argument) results for the transfer function H (left column) and the $\text{FFT}^{-1}\{H\}$ (right column). H is oversampled for (a) and (b); critically sampled for (c) and (d); and undersampled for (e) and (f). Solid lines are the discrete results; dashed lines are analytic curves. Phase is downward or upward parabolic curves. The gray "hash" in (f) is the rapidly oscillating magnitude.

$$U_2(x,y) = U_1(x,y) \otimes \mathfrak{F}^{-1}\{H(f_X, f_Y)\}. \tag{A.7}$$

Ideally, $\text{FFT}^{-1}\{H\}$ should be a sampled version of the impulse function h,

$$h(x,y) = \frac{e^{jkz}}{j\lambda z} \exp\left[\frac{jk}{2z}(x^2 + y^2)\right]. \tag{A.8}$$

Figure A.1(b) shows a profile of the magnitude and unwrapped phase for $\text{FFT}^{-1}\{H\}$ corresponding to Fig. A.1(a). Rather than a constant magnitude value, a window-like function appears. The phase result follows the analytic phase but limits at an absolute slope of π.

The phase result of Fig. A.1(b) is not of much consequence in this case, but the window-like magnitude requires some discussion. Goodman shows this window-like magnitude form is expected for the transform of a chirp function windowed with a rectangle function (see pp. 15–18 in *Introduction to Fourier*

Optics).³ The effective full width of the windowed magnitude form in Fig. A.1(b) is approximately

$$D_{FFT^{-1}\{\text{Oversampled } H\}} \approx \frac{\lambda z}{\Delta x}. \quad (A.9)$$

For a source field with support D_1 and considering the convolution of Eq. (A.7) with the result of Eq. (A.9), the width D_2 of the field in the observation plane that can be modeled accurately is approximately limited by

$$D_2 \leq D_1 + \frac{\lambda z}{\Delta x}. \quad (A.10)$$

Thus, for an accurate result, the field in the observation plane should fit within the width $D_1 + \lambda z/\Delta x$, which for oversampling is usually smaller than L. So oversampling H can limit the *support available in the observation plane*. For simple source functions, the most significant part of the field in the observation plane typically fits within $D_1 + \lambda z/\Delta x$; so, the limitation tends to affect the "wings" of the field.

A.1.2 Critically sampled transfer function

The transfer function is *critically* sampled when

$$\Delta x = \frac{\lambda z}{L}. \quad (A.11)$$

It is a remarkable property of sampled chirp functions that when Eq. (A.11) is true, the sampled FFT pair, H and h, exactly match the analytic pair H and h. Figures A.1(c) and (d) illustrate that the sampled H and $FFT^{-1}\{H\}$ exactly match the analytic functions. We might take the view that under ideal sampling conditions the periodicity of the sampled FFT pair, h and H, exactly matches the periodicity inherent in the FFT. Ideal sampling allows full use of the source and observation plane support and bandwidth allocation.

As might be expected, ideal sampling usually provides a simulation result that most closely *follows the analytic Fresnel result*.

A.1.3 Undersampled transfer function

The transfer function is *undersampled* when

$$\Delta x < \frac{\lambda z}{L}. \quad (A.12)$$

This is the relatively "long" distance propagation scenario. Figure A.1(e) shows the unwrapped argument of H for the case where $\Delta x = \lambda z/2L$. Undersampling results in a repetition of the fundamental phase profile where the periodic forms are actually comprised of aliased phase values. The "cusps" in the phase profile are characteristic of phase aliasing. The full spectral width where the phase is unaliased is given by

$$2B_{\text{Undersampled } H} = \frac{L}{\lambda z}. \qquad (A.13)$$

The aliased phase values can lead to significant artifacts in a simulation. This is suggested in Fig. A.1(f), which shows $\text{FFT}^{-1}\{H\}$. In Fig. A.1(f), the magnitude oscillates every other sample between a value of zero and $2/\lambda z$, which causes the gray appearance. In a simulation result this situation tends to cause "spiky" and "stair-step" artifacts in the observation plane field.

By avoiding the aliased part of the phase in a simulation, the spiky and stair-step artifacts can be reduced. So, for this sampling regime the following requirement is placed on the source bandwidth B_1:

$$B_1 \leq \frac{L}{2\lambda z}. \qquad (A.14)$$

Thus, undersampling H tends to *limit the spatial bandwidth that is available in the source plane*.

For ideal sampling, Eq. (A.14) becomes $B_1 \leq 1/2\Delta x$, which is simply the sampling theorem. Violating Eq. (A.14) also implies that significant parts of the observation field may not fit within the array side length. This can be argued by rearranging Eq. (A.14) to $2B_1\lambda z \leq L$ and recognizing that $2B_1\lambda z$ is the width (support) of the Fraunhofer pattern.

A.2 Fresnel Impulse Response Function Sampling

The Fresnel impulse response, repeated here,

$$h(x, y) = \frac{e^{jkz}}{j\lambda z} \exp\left[\frac{jk}{2z}\left(x^2 + y^2\right)\right], \qquad (A.15)$$

contains a spatial chirp function, and extracting the phase yields

$$\phi_h(x, y) = \frac{k}{2z}\left(x^2 + y^2\right). \qquad (A.16)$$

Proceeding in the same manner as for the transfer function, the criterion for adequately sampling the impulse response is

$$\Delta x \le \frac{\lambda z}{2|x_{max}|}. \quad (A.17)$$

Typically, $|x_{max}| = L/2$, so Eq. (A.17) becomes

$$\Delta x \le \frac{\lambda z}{L}. \quad (A.18)$$

The sense of Eqs. (A.5) and (A.18) is *exactly opposite*, and the two criteria are only satisfied simultaneously when $\Delta x = \lambda z/L$, which is the critical sampling condition.

A.2.1 Undersampled impulse response

The undersampled impulse response condition is

$$\Delta x > \frac{\lambda z}{L}, \quad (A.19)$$

which is the "short" distance scenario. The situation for h is shown in Fig. A.2(a). The magnitude of h is as expected (flat line) but the phase is aliased as evidenced by the cusps in the profile. The width between the phase cusps is

$$D_{\text{Undersampled } h} = \frac{\lambda z}{\Delta x}. \quad (A.20)$$

The FFT$\{h\}$, shown in Fig. A.2(b), displays the same trouble seen in Fig. A.1(f). Envisioning the IR propagation method as $h \otimes U_1$, then based on Eq. (A.20) the observation plane field will have a repeated form with an interval of $\lambda z/\Delta x$. So, in its own way the IR approach also limits the *support available in the observation plane* for the "short" distance scenario.

A.2.2 Critically sampled impulse response

For the critical sampling condition

$$\Delta x = \frac{\lambda z}{L}, \quad (A.21)$$

the sampled impulse response and its FFT are exactly the same as the analytic expression and exactly the same as the transfer function results [Figs. A.1(c) and (d)].

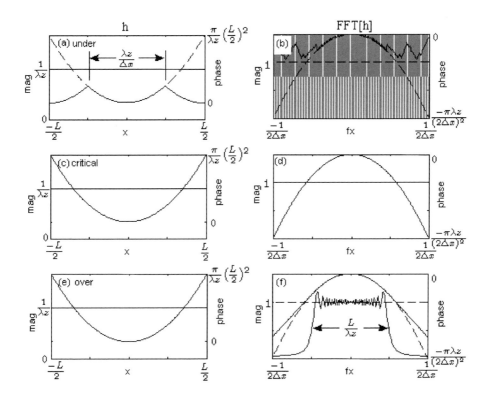

Figure A.2 Magnitude and phase (unwrapped argument) results for the impulse response h (left column) and FFT$\{h\}$ (right column). h is undersampled for (a) and (b); critically sampled for (c) and (d); and oversampled for (e) and (f). Solid lines are the discrete results; dashed lines are analytic curves. Phase is an upward or downward parabolic curve. The gray "hash" in (b) is the rapidly oscillating magnitude.

A.2.3 Oversampled impulse response

For the "long" distance scenario, the impulse response is oversampled:

$$\Delta x < \frac{\lambda z}{L}. \tag{A.22}$$

Here, the magnitude and phase of h follow the analytic form [Fig. A.2(e)] but FFT$\{h\}$ shows the windowed magnitude form and the limiting phase effect [Fig. A.2(f)]. The full width of the magnitude window in the frequency domain is

$$2B_{FFT\{\text{Oversampled } h\}} \approx \frac{L}{\lambda z}. \tag{A.23}$$

This window acts as a low-pass filter such that source frequency content greater than $\pm L/(2\lambda z)$ is significantly attenuated. Thus, the IR approach also *limits the spatial bandwidth that is available in the source plane*.

A.3 Summary

(a) For the "short distance" both the TF and IR approaches effectively limit the available support in the observation plane. The TF approach limitation is $D \leq \lambda z/\Delta x$. Beyond this limit, the TF approach primarily attenuates the field. The IR approach creates copies of the observation plane field. The TF result is usually preferred in this regime.

(b) For critical sampling the approaches yield identical results. The TF method requires one less FFT, so it is usually preferred.

(c) For the "long distance" case, the limit to the available bandwidth in the source plane is $B \leq L/\lambda z$ for both the TF and IR approaches. The IR approach primarily attenuates the source spectra beyond the limit, whereas the TF approach applies an aliased phase. The IR result is typically more usable in this case.

A.4 References

1. D. G. Voelz and M. C. Roggemann, "Digital simulation of scalar optical diffraction: Revisiting chirp function sampling criteria and consequences," *Appl. Opt.*, **48**, 6132–6142 (2009).

2. U. Spagnolini, "2-D phase unwrapping and phase aliasing," *Geophysics*, **58**, 1324–1334 (1993).

3. J. W. Goodman, *Introduction to Fourier Optics*, 3rd Ed., Roberts & Company, Greenwood Village, CO (2005).

Appendix B
Fresnel Two-Step Propagator

In some situations it is convenient for the source and observation plane side lengths to be different; for example, when a specific geometry in the source and observation planes must be maintained. A two-step propagation method developed in the 1980s and described more recently by others allows the side lengths to be chosen independently.[1-4]

B.1 Approach

The approach is derived by envisioning two artificial Fresnel propagations (Fig. B.1). The first is a propagation of the source field $U_1(x_1, y_1)$, a distance z_1 to a dummy plane, where $U_d(x_d, y_d)$ indicates the dummy plane field. The second is the propagation of the observation plane field $U_2(x_2, y_2)$ a distance z_2 to the dummy plane. Using Eq. (4.25), these propagations can be written as

$$U_d(x_d, y_d) = \frac{e^{jkz_1}}{j\lambda z_1} \exp\left[j\frac{k}{2z_1}\left(x_d^2 + y_d^2\right)\right]$$
$$\times \mathfrak{I}\left\{U_1(x_1, y_1)\exp\left[j\frac{k}{2z_1}\left(x_1^2 + y_1^2\right)\right]\right\}\Bigg|_{\substack{f_{X1} \to \frac{x_d}{\lambda z_1}, \\ f_{Y1} \to \frac{y_d}{\lambda z_1}}}$$

$$U_d(x_d, y_d) = \frac{e^{jkz_2}}{j\lambda z_2} \exp\left[j\frac{k}{2z_2}\left(x_d^2 + y_d^2\right)\right]$$
$$\times \mathfrak{I}\left\{U_2(x_2, y_2)\exp\left[j\frac{k}{2z_2}\left(x_2^2 + y_2^2\right)\right]\right\}\Bigg|_{\substack{f_{X2} \to \frac{x_d}{\lambda z_2}, \\ f_{Y2} \to \frac{y_d}{\lambda z_2}}}.$$

(B.1)

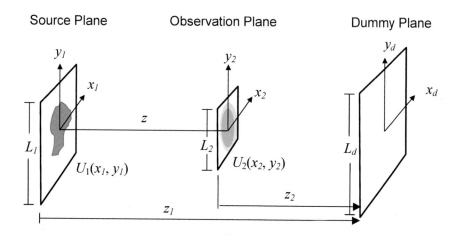

Figure B.1 Fresnel two-step propagation concept.

Equate the dummy plane fields and rearrange terms to find the field at the observation plane in terms of the source field,

$$U_2(x_2, y_2) = \frac{z_2}{z_1} \exp\left[jk(z_1 - z_2)\right] \exp\left[-j\frac{k}{2z_2}(x_2^2 + y_2^2)\right]$$
$$\times \mathfrak{I}^{-1}\left\{ \begin{array}{l} \exp\left[j\frac{k}{2}\left(\frac{1}{z_1} - \frac{1}{z_2}\right)(x_d^2 + y_d^2)\right] \\ \times \mathfrak{I}\left\{U_1(x_1, y_1) \exp\left[j\frac{k}{2z_1}(x_1^2 + y_1^2)\right]\right\} \end{array} \right\}.$$

(B.2)

The propagation distance from the source to the observation plane is given by

$$z = z_1 - z_2.$$ (B.3)

The dummy plane coordinates are related to the transform frequencies in the source and observation planes by

$$x_d = \lambda z_1 f_{X1} = \lambda z_2 f_{X2}.$$ (B.4)

Only the x dimension is indicated here, but y-dimension parameters follow similarly. With discrete sampling, the source and observation plane side lengths are given by

$$L_1 = M\Delta x_1, \qquad L_2 = M\Delta x_2,$$ (B.5)

Fresnel Two-Step Propagator

where M is the number of samples (linear) and Δx_1 and Δx_2 are the respective sample intervals. In frequency space $f_{X1} = p/2\Delta x_1$ and $f_{X2} = p/2\Delta x_2$, where p is an index ranging from $-M/2$ to $M/2-1$. With these definitions and applying Eq. (B.4) we find

$$\frac{z_1}{z_2} = \frac{L_1}{L_2} = \frac{\Delta x_1}{\Delta x_2}. \tag{B.6}$$

With the help of Eq. (B.3), the following are derived:

$$z_1 = z\left(\frac{L_1}{L_1 - L_2}\right); \quad z_2 = z\left(\frac{L_2}{L_1 - L_2}\right). \tag{B.7}$$

It is acceptable for $L_2 > L_1$, where z_1 and z_2 are negative and imply the dummy plane is located to the "left" of the source and observation planes. Substitute $x_d = \lambda z_1 f_{X1}$, $y_d = \lambda z_1 f_{Y1}$, and Eq. (B.7) into Eq. (B.2), and perform the algebra necessary to find

$$U_2(x_2, y_2) = \frac{L_2}{L_1} \exp(jkz) \exp\left[-j\frac{k}{2z}\frac{(L_1 - L_2)}{L_2}(x_2^2 + y_2^2)\right]$$
$$\times \mathfrak{F}^{-1}\left\{\begin{array}{l}\exp\left[-j\pi\lambda z \frac{L_1}{L_2}(f_{X1}^2 + f_{Y1}^2)\right] \\ \times \mathfrak{F}\left\{U_1(y_1, x_1) \exp\left[j\frac{k}{2z}\frac{(L_1 - L_2)}{L_1}(x_1^2 + y_1^2)\right]\right\}\end{array}\right\}. \tag{B.8}$$

The expression in Eq. (B.8) is the basis for the two-step propagator, where the source and observation plane side lengths, L_1 and L_2, respectively, can be chosen independently. In a simulation, the source and observation plane coordinates are defined in the usual way:

$$x_1 \to [-L_1/2 : \Delta x_1 : L_1/2 - \Delta x_1],$$
$$x_2 \to [-L_2/2 : \Delta x_2 : L_2/2 - \Delta x_2]. \tag{B.9}$$

Examining Eq. (B.8), it is apparent that when $L_1 = L_2$ the two-step method collapses to the transfer function method described in Section 5.1.

B.2 Sampling Considerations

Sampling considerations for the two-step method can be approached by first defining the oversampled regimes for the three chirp functions in Eq. (B.8). The source chirp given by

$$\exp\left[j\frac{k}{2z}\frac{(L_1-L_2)}{L_1}\left(x_1^2+y_1^2\right)\right]$$

is oversampled when $\Delta x_1 \leq \lambda z_1/L_1$ (see Appendix A). Using the relation for z_1 in Eq. (B.7) yields

$$\Delta x_1 \leq \frac{\lambda z}{|L_1-L_2|}. \qquad (B.10)$$

The absolute value accounts for the case where $L_2 > L_1$. For the observation plane chirp,

$$\exp\left[-j\frac{k}{2z}\frac{(L_1-L_2)}{L_2}\left(x_2^2+y_2^2\right)\right],$$

the requirement for oversampling is $\Delta x_2 \leq \lambda z_2/L_2$, which leads to $\Delta x_2 \leq \lambda z/|L_1-L_2|$. To allow for comparison in terms of the source plane sampling, apply Eq. (B.6) to get

$$\Delta x_1 \leq \frac{L_1}{L_2}\frac{\lambda z}{|L_1-L_2|}. \qquad (B.11)$$

For the frequency domain chirp,

$$\exp\left[-j\pi\lambda z\frac{L_1}{L_2}\left(f_{X1}^2+f_{Y1}^2\right)\right],$$

following arguments that led to Eq. (A.5), this chirp is oversampled when

$$\Delta x_1 \geq \frac{\lambda z}{L_2}. \qquad (B.12)$$

For critical sampling, Eqs. (B.10)–(B.12) become equalities. Equating Eqs. (B.10) and (B.11) results in the condition $L_2 = L_1$. Thus, critical sampling requires the side lengths to be equal. Also, under critical sampling, Eq. (B.12) gives the familiar critical sampling criterion $\Delta x_1 \equiv \lambda z / L_1$. As noted previously, with $L_2 = L_1$ the two-step method is identical to the transfer function (TF) approach.

The utility of the two-step approach is that L_1 and L_2 can be different sizes, although this requires noncritical sampling. Sampling criteria involving the source bandwidth or support are complicated to develop for this method, as there are a number of possible combinations depending on simulation parameter choices. A few examples follow.

B.2.1 Similar side lengths

A common simulation situation is where the side length difference $|L_1 - L_2|$ is relatively small. This arrangement may be used, for example, when modeling a propagating field that is slowly expanding. Under this circumstance the source and observation chirps are likely to be oversampled [i.e., Eqs. (B.10) and (B.12) are true].

For the frequency chirp, consider the right side of Eq. (B.8) where the product of the source field U_1 and source chirp function are Fourier transformed. This results in a convolution of the spectra of these two quantities. The approximate full width of this transform spectrum is the spectral width of U_1, which is $2B_1$, plus the spectral width of the source chirp function. The source chirp full spectral width is found to be $\sim (L_2-L_1)/(\lambda z)$ [see Eq. (A.13)]. Furthermore, if the frequency domain chirp is undersampled, it will have an unaliased full width of $\sim L_2/(\lambda z)$. The sampling criterion is found by recognizing that the width of the transform result should "fit" within the unaliased frequency chirp width. This is stated as

$$2B_1 + \left(\frac{L_2 - L_1}{\lambda z}\right) \leq \frac{L_2}{\lambda z}. \quad (B.13)$$

Rearranging Eq. (B.13) yields

$$B_1 \leq \frac{L_1}{2\lambda z}, \quad (B.14)$$

which is identical to the *oversampling condition for the transfer function* method (Table 5.1). To summarize, when L_1 and L_2 are close to the same size, the two-step sampling criterion is essentially the same as the TF approach. The size of the observation plane L_2 is not a factor in the criterion. L_2 affects the observation plane chirp, but this term is oversampled and therefore not of concern.

B.2.2 Significantly different side lengths

When L_1 and L_2 differ significantly, by a factor of 2 or more, the source chirp can have an aliasing problem. The center support region in which the chirp phase is unaliased is $\sim \lambda z/\Delta x_1$ [see Eq. (A.12)]. For the source to avoid the aliased region, the following is necessary:

$$D_1 \leq \frac{\lambda z}{\Delta x_1}, \qquad (B.15)$$

where D_1 is the source support. Further trouble can be caused by frequency chirp aliasing, but the specific effect depends on the ratio L_1/L_2.

B.2.3 Comments and recommendations

(a) The two-step method reduces to the transfer function for critical sampling including the requirement $L_1 = L_2$.

(b) Choosing L_1 and L_2 to be different forces noncritical sampling. Sampling criteria depend on the relationship between L_1 and L_2.

(c) The two-step method suffers the same type of artifacts as the TF approach for longer distances; however, replacing the frequency chirp term with a "windowed" chirp term allows performance similar to the impulse response (IR) approach.[5]

(d) The implication of the sampling investigation is that the two-step method does not alleviate the sampling constraints discussed in Chapter 5 for the Fresnel propagators. It provides a method to resample the observation plane grid within the propagation process.

(e) Experience suggests working close to the critical sampling regime ($L_1 \approx L_2$) results in the least artifacts. If $L_2 > L_1$ (for example, more than a factor of 2 or 3) the result tends to pick up sidelobes in the wings of the pattern. If $L_2 < L_1$ the pattern tends to pick up spiky artifacts or oscillations.

B.3 MATLAB Code

The expression in Eq. (B.8) is coded in the following function `prop2step`:

```
1  function [u2]=prop2step(u1,L1,L2,lambda,z)
2  % propagation - 2 step Fresnel diffraction method
3  % assumes uniform sampling and square array
4  % u1 - complex field at source plane
5  % L1 - source plane side-length
6  % L2 - observation plane side-length
7  % lambda - wavelength
8  % z - propagation distance
9  % u2 - output field at observation plane
10 %
11 [M,N]=size(u1);          %input array size
12 k=2*pi/lambda;           %wavenumber
```

```
13
14  % source plane
15  dx1=L1/M;
16  x1=-L1/2:dx1:L1/2-dx1;
17  [X,Y]=meshgrid(x1,x1);
18  u=u1.*exp(j*k/(2*z*L1)*(L1-L2)*(X.^2+Y.^2));
19  u=fft2(fftshift(u));
20
21  % dummy (frequency) plane
22  fx1=-1/(2*dx1):1/L1:1/(2*dx1)-1/L1;
23  fx1=fftshift(fx1);
24  [FX1,FY1]=meshgrid(fx1,fx1);
25  u=exp(-j*pi*lambda*z*L1/L2*(FX1.^2+FY1.^2)).*u;
26  u=ifftshift(ifft2(u));
27
28  % observation plane
29  dx2=L2/M;
30  x2=-L2/2:dx2:L2/2-dx2;
31  [X,Y]=meshgrid(x2,x2);
32  u2=(L2/L1)*u.*exp(-j*k/(2*z*L2)*(L1-L2)*(X.^2+Y.^2));
33  u2=u2*dx1^2/dx2^2;    %x1 to x2 scale adjustment
34  end
```

B.4 References

1. G. A. Tyler and D. L. Fried, "A wave optics propagation algorithm," *Report TR-451,* Optical Sciences Company, Anaheim, CA (1982).

2. C. Rydberg and J. Bengtsson, "Efficient numerical representation of the optical field for the propagation of partially coherent radiation with a specified spatial and temporal coherence function," *J. Opt. Soc. Am. A,* **23,** 1616–1625 (2006).

3. X. Deng, B. Bihari, J. Gan, F. Zhao, and R. T. Chen, "Fast algorithm for chirp transforms with zooming-in ability and its applications," *J. Opt. Soc. Am. A,* **17,** 762–771 (2000).

4. S. Coy, "Choosing mesh spacings and mesh dimensions for wave optics simulation," *Proc. SPIE* **5894,** 589405 (2005). doi:[10.1117/12.619994]

5. D. G. Voelz and M. C. Roggemann, "Digital simulation of scalar optical diffraction: Revisiting chirp function sampling criteria and consequences," *Appl. Opt.,* **48,** 6132–6142 (2009).

Appendix C
MATLAB Function Listings

C.1 Circle

```
function[out]=circ(r);
%
% circle function
%
% evaluates circ(r)
% note: returns odd number of samples for diameter
%
out=abs(r)<=1;
end
```

C.2 Jinc

```
function[out]=jinc(x);
%
% jinc function
%
% evaluates J1(2*pi*x)/x
% with divide by zero fix
%
% locate non-zero elements of x
mask=(x~=0);
% initialize output with pi (value for x=0)
out=pi*ones(size(x));
% compute output values for all other x
out(mask)=besselj(1,2*pi*x(mask))./(x(mask));
end
```

C.3 Rectangle

```
function[out]=rect(x);
%
% rectangle function
%
% evaluates rect(x)
```

```
% note: returns odd number of samples for full width
%
out=abs(x)<=1/2;
end
```

C.4 Triangle

```
function[out]=tri(x)
%
% triangle function
%
% evaluates tri(x)
%
% create lines
t=1-abs(x);
% keep lines for |x|<=1, out=0 otherwise
mask=abs(x)<=1;
out=t.*mask;
end
```

C.5 Unit Sample "Comb"

```
function[out]=ucomb(x);
%
% unit sample "comb" function
%
% sequence of unit values for x=integer value
% round is used to truncate roundoff error
%
x=round(x*10^6)/10^6;    %round to 10^6ths place
out=rem(x,1)==0;         %place 1 in out where rem = 0
end
```

C.6 Unit Sample "Delta"

```
function[out]=udelta(x);
%
% unit sample "delta" function
%
% unit value for x=0
% round is used to truncate roundoff error
%
x=round(x*10^6)/10^6;    %round to 10^6ths place
out=x==0;                %place 1 in out where x = 0
end
```

Appendix D
Exercise Answers and Results

D.1 Chapter 1

Exercise 1.1

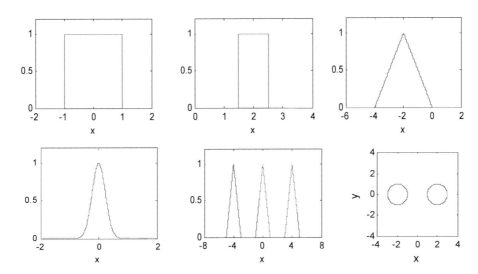

Exercise 1.2

(a) $4w^2 \operatorname{sinc}(2wf_X)\operatorname{sinc}(2wf_Y)$;

(b) $4w^2 \operatorname{sinc}(2wf_X)\operatorname{sinc}(2wf_Y)\exp(-j2\pi x_0 f_X)$;

(c) $\pi w^2 \exp\left[-\pi^2 w^2 \left(f_X^2 + f_Y^2\right)\right]$;

(d) $w_1^2 \dfrac{J_1\left(2\pi w_1 \sqrt{f_X^2 + f_Y^2}\right)}{w_1 \sqrt{f_X^2 + f_Y^2}} - w_2^2 \dfrac{J_1\left(2\pi w_2 \sqrt{f_X^2 + f_Y^2}\right)}{w_2 \sqrt{f_X^2 + f_Y^2}}$;

(e) $w^2 \dfrac{J_1\left(2\pi w \sqrt{f_X^2 + f_Y^2}\right)}{w \sqrt{f_X^2 + f_Y^2}} \cos(\pi d f_X)$.

209

Exercise 1.3

(a) $4w^2 \Lambda\left(\dfrac{x}{2w}\right)\Lambda\left(\dfrac{y}{2w}\right)$;

(b) $\dfrac{144}{25}\exp\left(-\pi\dfrac{x^2+y^2}{5^2}\right)$;

(c) $2\,\text{sinc}\left(\dfrac{x}{4}\right)\text{sinc}(y)$.

Exercise 1.4

(a) $4w^2 \Lambda\left(\dfrac{x}{2w}\right)\Lambda\left(\dfrac{y}{2w}\right)$;

(b) $\dfrac{w^2}{2}\exp\left(-\pi\dfrac{x^2+y^2}{2w^2}\right)$.

Exercise 1.5

(a) $8w^2$;

(b) 9π.

Exercise 1.6

(a) Linear, space-invariant.
(b) Nonlinear, this is an affine transformation, but space-invariant.
(c) Nonlinear, space-invariant.
(d) Linear, not space-invariant.
(e) Linear, space-invariant.

D.2 Chapter 2

Exercise 2.1

Sample number: 500
Nyquist frequency: 5×10^4 cycles/m.
Frequency sample interval: 200 cycles/m;
Range of coordinates in the spatial domain: [−2.5, 2.49 mm];
Range of coordinates in the frequency domain: [-5×10^4, 4.98×10^4 cycles/m].

Exercise 2.2

(a) 5 cycles/mm, 0.1 mm, 25.6 mm;
(b) 0.446 cycles/mm; 1.12 mm; 287 mm.

Exercise 2.3

$2.42w$.

Exercise Answers and Results

Exercise 2.4

    (a) $1/(2w)$;

    (b) $1/w$.

Exercise 2.5

    (a) $6d$;

    (b) $4d$.

D.3 Chapter 3

Exercise 3.1

    (e) $w\,\mathrm{sinc}^2(wf_X)$.

    (f) Phase plot should be zero for all frequencies, but finite calculation precision results in jumps at zero magnitude positions.

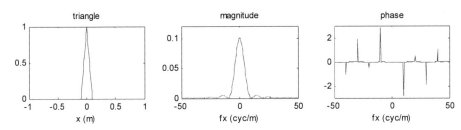

Exercise 3.2

$$\frac{w_a w_b}{\sqrt{w_a^2 + w_b^2}} \exp\left(-\pi \frac{x^2}{w_a^2 + w_b^2}\right).$$

Exercise 3.3

    (a)–(d):

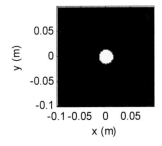 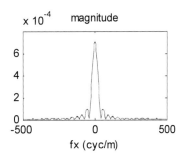

D.4 Chapter 4

Exercise 4.1

$$\text{OPD} = d_1(n_1 - 1) - d_2(n_2 - 1).$$

Exercise 4.2

RS, $z \gg \lambda$; using N_F criteria: Fresnel $z > \sim 0.5$ m, Fraunhofer $z > \sim 5$ m.

Exercise 4.4

(a) $U(x,y) = \dfrac{e^{jkz} e^{j\frac{k}{2z}(x^2+y^2)}}{j\lambda z} 4 w_X w_Y \,\text{sinc}\left(\dfrac{2w_X}{\lambda z}x\right)\text{sinc}\left(\dfrac{2w_Y}{\lambda z}y\right);$

$I(x,y) = \left(\dfrac{4w_X w_Y}{\lambda z}\right)^2 \text{sinc}^2\left(\dfrac{2w_X}{\lambda z}x\right)\text{sinc}^2\left(\dfrac{2w_Y}{\lambda z}y\right).$

$z = 5$ m; $L = 0.2$ m

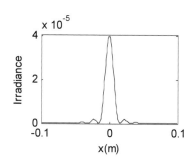

(b) $U(x,y) = \dfrac{e^{jkz} e^{j\frac{k}{2z}r^2}}{j\lambda z}\left[w_1^2 \dfrac{J_1\left(2\pi\dfrac{w_1}{\lambda z}r\right)}{\dfrac{w_1}{\lambda z}r} - w_2^2 \dfrac{J_1\left(2\pi\dfrac{w_2}{\lambda z}r\right)}{\dfrac{w_2}{\lambda z}r}\right],$

$r = \sqrt{x^2 + y^2}\,;$

$I(x,y) = \left(\dfrac{1}{\lambda z}\right)^2 \left[w_1^2 \dfrac{J_1\left(2\pi\dfrac{w_1}{\lambda z}r\right)}{\dfrac{w_1}{\lambda z}r} - w_2^2 \dfrac{J_1\left(2\pi\dfrac{w_2}{\lambda z}r\right)}{\dfrac{w_2}{\lambda z}r}\right]^2.$

$z = 50$ m; $L = 0.2$ m

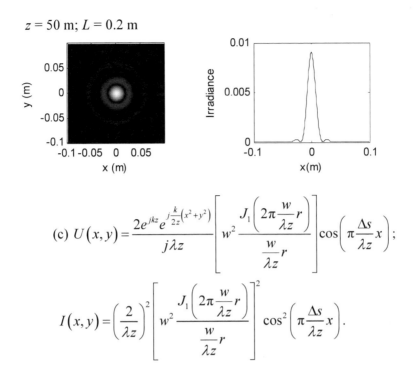

(c) $U(x,y) = \dfrac{2e^{jkz} e^{j\frac{k}{2z}(x^2+y^2)}}{j\lambda z} \left[w^2 \dfrac{J_1\left(2\pi \dfrac{w}{\lambda z} r\right)}{\dfrac{w}{\lambda z} r} \right] \cos\left(\pi \dfrac{\Delta s}{\lambda z} x\right);$

$I(x,y) = \left(\dfrac{2}{\lambda z}\right)^2 \left[w^2 \dfrac{J_1\left(2\pi \dfrac{w}{\lambda z} r\right)}{\dfrac{w}{\lambda z} r} \right]^2 \cos^2\left(\pi \dfrac{\Delta s}{\lambda z} x\right).$

$z = 50$ m; $L = 0.1$ m

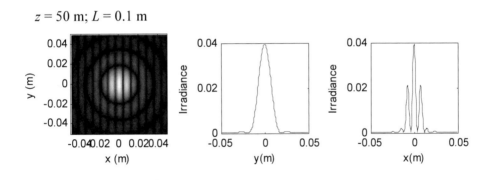

Exercise 4.5

$I(x,y) = \dfrac{A_1^2 + A_2^2}{(\lambda z)^2} \left[w^2 \dfrac{J_1\left(2\pi \dfrac{w}{\lambda z} r\right)}{\dfrac{w}{\lambda z} r} \right]^2 \left[1 + \dfrac{2 A_1 A_2}{A_1^2 + A_2^2} \cos\left(2\pi \dfrac{\Delta s}{\lambda z} x\right) \right].$

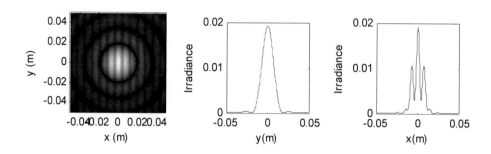

D.5 Chapter 5

Exercise 5.1

(a) 0.5 m, 0.001 m, 500 cycles/m.
(b) 100 cycles/m, yes, 100.
(c) 5, okay for this simple aperture.
(d) Axes scaled for display.

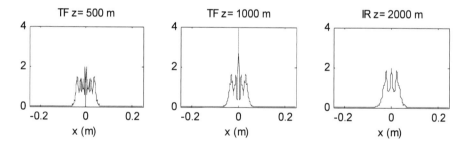

Exercise 5.2

(a) Simulation for $L_1 = 2$ mm. Axes scaled for display.

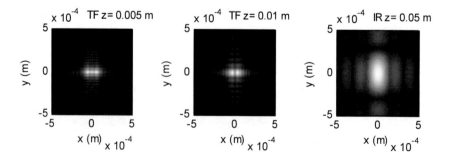

(b) Simulation with $L_1 = 2.5$ cm. Axes scaled for display.

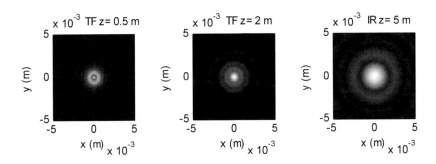

(c) Simulation run with $L_1 = 2.5$ cm. Axes scaled for display.

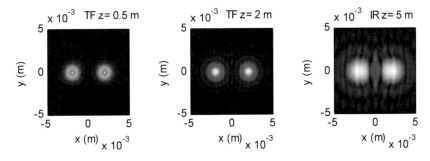

z for critical sampling: (a) 1.26 cm; (b) 1.97 m; (c) 1.97 m.

Exercise 5.3

(a) $P = 0.0104$.
(b) Discrepancies for short- and long-distance IR results.

Exercise 5.4

(b) Subtle differences.
(c) 5.2, 2.6, 1.3, 0.26. The Fresnel propagator results compare well with the Rayleigh–Sommerfeld results for all ranges.

Exercise 5.5

(a)

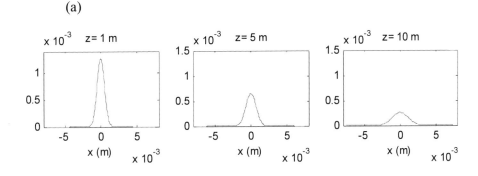

(b) 1.42 m; B_1 = 445.7 cycles/m < 1185 cycles/m—criterion is met.

Exercise 5.6

Results should be the same.

Exercise 5.7

(a)

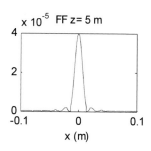

(b)

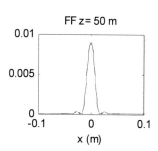

(c)

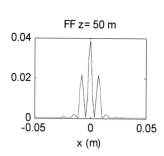

Exercise 5.8

$M \gg 100$

Exercise 5.9

(a)

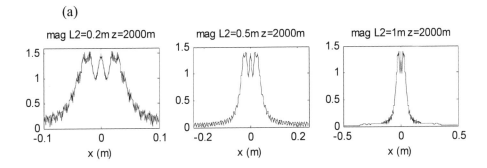

$L_2 = 0.2$ m—spiky features; $L_2 = 1$ m—loss of oscillations in the wings.

D.6 Chapter 6

Exercise 6.2

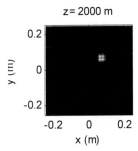

Exercise 6.3

(b)

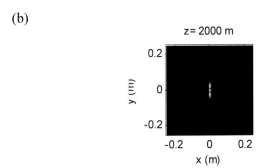

(c) No, the Fresnel propagation chirp is canceled only in one axis. However, for a simulation the single FFT propagation [Eq. (4.25)] can be used.

Exercise 6.5

(a) Creates both converging and diverging wavefronts.
(b) $\Delta r \leq \lambda f/(2w)$.
(c) $\Delta x = 0.1$ mm, $\lambda f/(2w) = 0.4$ mm—criterion is satisfied.
(d) Critical sampling (use TF).
(e)

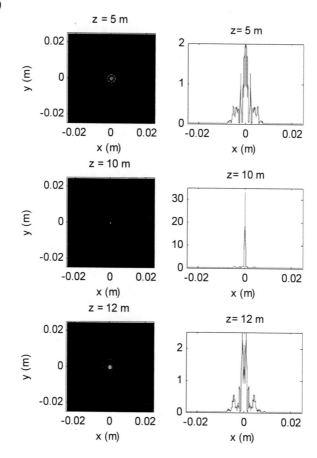

(g) and (h) x axis is scaled for display.

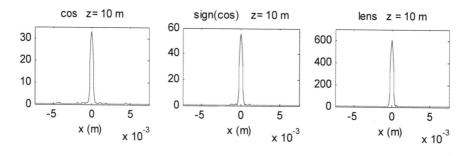

The sign(cos) zone plate puts more irradiance on axis because more light is transmitted overall through the plate. The diffraction limited lens has a peak irradiance ~ 10 times larger than the sign(cos) zone plate.

Exercise 6.6

To sample the periodic square wave consistently requires $M = i \cdot L_1/P = i \cdot 100$, where i is an integer. But to also get an odd number of samples across the rect requires $M = 2(1 + 2i) \cdot L_1/P$. So, the values of M that work are 200, 600, 1000, 1400, etc. Therefore, 1400 is the next value above 1000.

Exercise 6.7

Examples:

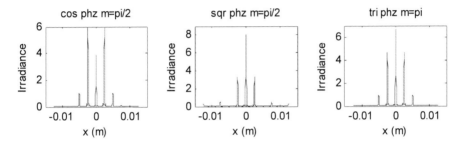

Exercise 6.8

(a) One set of acceptable parameters is $L_1 = 0.8 \times 10^{-3}$ and $M = 3200$. From Eq. (6.21) this gives eight samples across P.

(b)
$$U_2(x_2) = \frac{\exp(jkf)}{\sqrt{j\lambda f}} \exp\left[j\frac{k}{2f}x_2^2\right] \frac{D_1}{2} \sum_{n=-\infty}^{\infty} \mathrm{sinc}^2\left(\frac{n}{2}\right) \mathrm{sinc}\left[\frac{D_1}{\lambda f}\left(x_2 - n\frac{\lambda f}{P}\right)\right],$$

and $I_2(x_2) = |U_2(x_2)|^2$

(c)

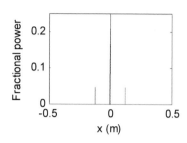

D.7 Chapter 7

Exercise 7.1

(a) 66.7 mm, −0.33.
(b) 20 mm, 66.7 mm, 3.33.
(c) $z_2 \approx f$.

Exercise 7.2

(a) 8, 106 cycles/mm, 213 cycles/mm.
(b) $\Delta u \leq 2.3$ µm.
(c) $M \geq 426$.

Exercise 7.3

(a) $f_{ox} = 40$ cycles/mm; $f_{oy} = 20$ cycles/mm; $2f_{ox} = 80$ cycles/mm; $2f_{oy} = 40$ cycles/mm.

(b)

(c)

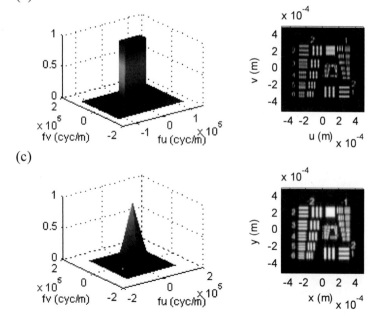

(d) Less resolution is observed in the horizontal bars than the vertical bars.

Exercise 7.4

(a) $f_o = 40$ cycles/mm; $2f_o = 80$ cycles/mm.

(b)

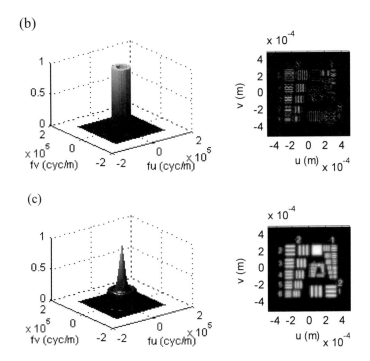

(c)

Exercise 7.5

(a) Smaller phase variance → less speckle contrast.
(b) Smaller aperture → larger speckle lobes.

Exercise 7.6

(a) $h(r) = f_0^2 \dfrac{J_1(2\pi f_0 r)}{f_0 r}$, $r = \sqrt{u^2 + v^2}$.

(b) $|h(r)|^2$.

Exercise 7.7

(a) 12.2 μm.
(b) 1.2 μm; $S \approx 10$.

(d) Dip ~ 20%.

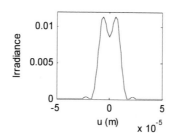

Exercise 7.8

(a) 7×10^5.
(b) 2.86 cycles/m; 0.43 m.
(c) 0.175 m; 179.2 m.

Exercise 7.9

(b)

(c)

(d)

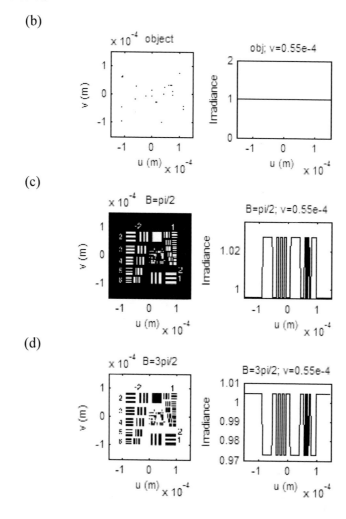

Exercise Answers and Results 223

The approximate analytic profiles have slight bias relative to the simulation profiles. The overall average image irradiance should be 1, which is true for the simulation result but not true for the analytic expressions.

Exercise 7.10

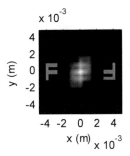

D.8 Chapter 8

Exercise 8.1

(a) 363 cycles/mm; 512 cycles/mm; yes.

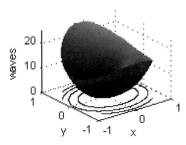

Exercise 8.2

(b) $W_d \approx -W_{040}$ gives smallest peak-to-peak value for W.

Exercise 8.3

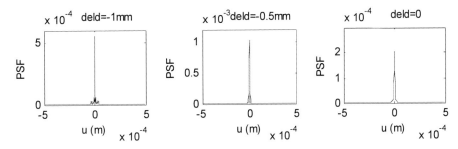

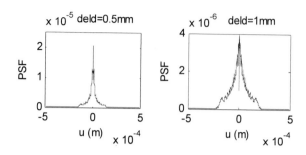

Peak PSF value corresponds to $\delta_d = -0.5$ mm, therefore, $W_d = -4.5\lambda$. This value roughly "balances" the initial spherical aberration for the on-axis image point ($W_{040} = 4.96\lambda$).

Exercise 8.4

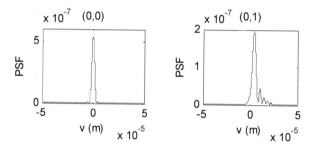

Exercise 8.5

$W_d = 19.4\lambda$. When the OPD function is undersampled, for example, $W_d = 25\lambda$—the resulting PSF extends beyond the array boundaries.

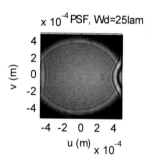

Exercise 8.6

Similar to Seidel results.

Exercise Answers and Results

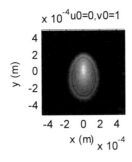
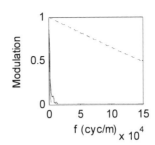

Exercise 8.7

(a)   (d)   (f)

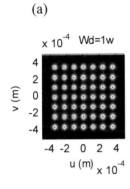
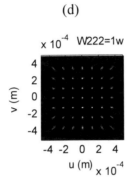
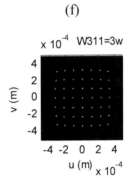

Exercise 8.8

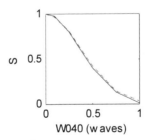

(e) $S = 0.0212$ for on-axis point.

D.9 Chapter 9

Exercise 9.1

(a)

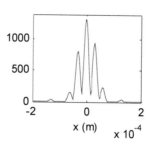

(b) $|\text{sinc}(\Delta v \cdot \Delta d/c)|$

Exercise 9.2

(b) range of 1λ.

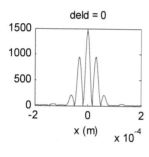
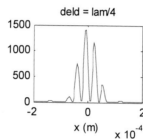
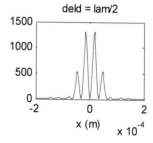
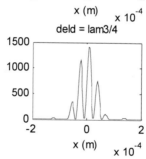

Exercise 9.3

(a) $V = 0.335$.

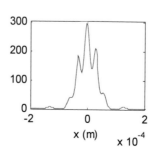

Exercise 9.4

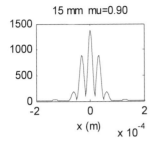
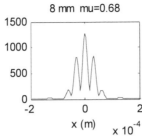
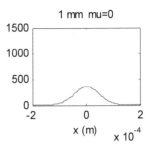

Exercise 9.5

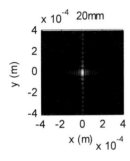
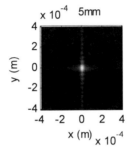
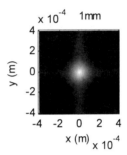

Exercise 9.6

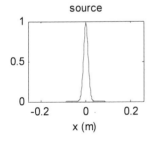
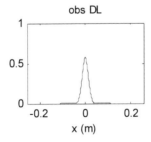
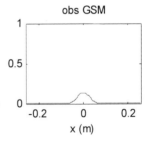

Index

aberrations, 141
abs, 36
Airy pattern, 59, 97, 109, 134
aliasing, 15, 24, 71
angle, 36
aperture stop, 114
astigmatism, 143
autocorrelation theorem, 4
axis, 33
axis square, 42
axis xy, 42

bandlimited function, 14
Bessel function, 7
besselj, 58

centered arrangement, 21
central ordinate theorem, 4
chirp function, 7, 191
chromatic aberration, 141
circ, 207
circle function, 5, 45, 207
circshift, 159
coh_image, 120
coherence, 48, 169
coherence length, 171
coherence time, 171
coherent cutoff frequency, 118
coherent image transfer function, 117, 147
coherent imaging, 116
colormap, 42
coma, 143
comb function, 5
command history, 29

command window, 29
complex coherence factor, 177, 182
complex degree of temporal coherence, 171, 176
conv_example, 40
converging wavefront, 93
convolution integral, 9
convolution theorem, 4
coordinates, 20
critically sampled condition, 73, 194, 196
cross-correlation theorem, 4
cross-spectrally pure, 186
current directory, 29
cylindrical lens, 108

defocus, 144
diffraction, 47
diffraction limited, 117
discrete Fourier transform (DFT), 18
distortion, 143
diverging wavefront, 93

editor window, 30
effective bandwidth, 15
entrance pupil, 114
exit pupil, 114

fast Fourier transform (FFT), 18
fft, 45
fft_example, 32
fft2, 45
fft2_example, 41
fftshift, 34

field curvature, 143
field of view, 149
figure, 32
flipud, 121
f-number ($f/\#$), 96, 115
focal length, 96, 113
focal plane, 97
focal ratio, 96
focus, 93, 94
Fourier integral theorem, 4
Fourier transform, 3
Fourier transform, inverse, 3
fractional image height, 142
fraun_circ, 58
Fraunhofer diffraction, 55
Fraunhofer region, 56
free space optical propagation, 48
Fresnel diffraction, 53
Fresnel impulse response propagator, 64, 195
Fresnel integrals, 75
Fresnel number, 55, 56
Fresnel propagators, 191
Fresnel transfer function propagator, 63, 191
Fresnel two-step propagator, 79, 199
fringe shift, 187
function, 30
functions, list of, 5

Gaussian beam, 85
Gaussian lineshape, 170
Gaussian Schell-model (GSM) beam, 188
Gaussian function, 5
geometrical optics, 2, 113
grating, 98
grating_cos, 99
grating_sqr, 103

holography, 137
Huygens–Fresnel principle, 52

ifft, 45
ifft2, 45

ifftshift, 36
image quality, 147
image_super, 160
imagesc, 42
impulse response, 8
imread, 121
incoh_image, 129
incoherent cutoff frequency, 129, 147
incoherent imaging, 127
incoherent light, 127
iris, 115
irradiance, 49

jinc function, 57, 207

leading and lagging phase, 51
lens, 96
lens_psfmtf, 149
lens law, 113
linear system, 7, 11
linearity theorem, 4
lineshape, 170
linewidth, 170

meshgrid, 41
M-files, 30
M-Lint, 83
modulation depth, 148
modulation transfer function, 148
monochromatic light, 48, 170

NaN, 145
nthroot, 44
Nyquist frequency, 15

object space, 135
optical path difference (OPD), 59, 141
optical path length (OPL), 50
optical transfer function (OTF), 127, 147
oversampled condition, 71, 192, 197

parabolic mirror, 164

paraxial working $f/\#$, 115
Parseval's (Rayleigh's) theorem, 4
partial spatial coherent, 169
partial temporal coherent, 169
pc_spatial, 182
pc_temp, 174
periodic convolution, 24
periodic extension, 21
phase contrast imaging, 136
phase grating, 110
phase screen, 178, 184, 187
phasor, 49
plane wave, 48
plot, 32
point spread function (PSF), 127, 148
polychromatic light, 170
power spectral density, 170
primary aberrations, 142
principal plane, 113
profiler, 83
prop2step, 204
propFF, 80
propIR, 65
propTF, 63
PSF image plane map, 157
psf_map, 157
pupil function, 96, 146

quasi-monochromatic light, 170

rand, 126
Rayleigh resolution criterion, 135
Rayleigh–Sommerfeld diffraction, 51
rect, 31, 207
rectangle function, 5, 207
rectangular lineshape, 187
reducible, 186
refractive index, 50
rough object, 124

sagittal plane, 153
sample interval, frequency, 19
sample interval, spatial, 13

sample rate, 13
sampling regimes, 73
sampling theorem, 14
scalar diffraction, 47
script, 30
Seidel polynomials, 142
seidel_5, 144
separable function, 3
shift theorem, 4
shifted arrangement, 21
side length, 14
similarity theorem, 4
sinc, 38
sinc function, 5
single, 121
space-invariant system, 7, 11
speckle, 126
spherical aberration, 143
split-step simulation, 78
sqr_beam, 66
Strehl ratio, 167
successive transform theorem, 4
sum, 84
superposition integral, 8, 160
support, 14
surf, 43

tangential plane, 153
test chart, 120
theorems, list of, 4
thin lens, 113
tilt, 89, 90
transfer function, 9
transmittance function, 89, 178
transverse coherence length, 178
transverse magnification, 114
tri, 208
triangle function, 5, 45, 208

ucomb, 208
udelta, 208
undersampled condition, 74, 194, 196
uniform sampling, 20
unit sample "comb" function, 208

unit sample "delta" function, 208
unwrap, 68

vignetting, 98
visibility, 187

wave optics, 2
wavelength, 49
wavenumber, 48

xlabel, 33

ZEMAX, 148
Zernike polynomials, 142, 165
zone plate, 109

 David G. Voelz is a professor of electrical engineering at New Mexico State University and holds a Paul W. and Valerie Klipsch Professorship. He received the B.S. degree in electrical engineering from New Mexico State University in 1981 and received the M.S. and Ph.D. degrees in electrical engineering from the University of Illinois in 1983 and 1987, respectively. From 1986 to 2001, he was with the Air Force Research Laboratory in Albuquerque, NM. He was named a Fellow of SPIE in 1999 and has received an OSA Engineering Excellence Award, the Bromilow Award at NMSU for research excellence, and the Giller Award at AFRL for technical achievement. His research interests always seem to involve some aspect of Fourier optics and include spectral and polarimetric imaging, laser imaging and beam projection, laser communication, adaptive optics, and astronomical instrumentation development.